CRAFT, COMMUNITY AND THE MATERIAL CULTURE OF PLACE AND POLITICS, 19TH–20TH CENTURY

Craft practice has a rich history and remains vibrant, sustaining communities while negotiating cultures within local or international contexts. More than two centuries of industrialization have not extinguished handmade goods; rat=her, the broader force of industrialization has redefined and continues to define the context of creation, deployment and use of craft objects. With object study at the core, this book brings together a collection of essays that address the past and present of craft production, its use and meaning within a range of community settings from the Huron Wendat of colonial Quebec to the Girls' Friendly Society of twentieth-century England.

The making of handcrafted objects has and continues to flourish despite the powerful juggernaut of global industrialization, whether inspired by a calculated refutation of industrial sameness, an essential means to sustain a cultural community under threat, or a rejection of the imposed definitions by a dominant culture. The broader effects of urbanizing, imperial and globalizing projects shape the multiple contexts of interaction and resistance that can define craft ventures through place and time. By attending to the political histories of craft objects and their makers, over the last few centuries, these essays reveal the creative persistence of various hand mediums and the material debates they represented.

Janice Helland is Professor of Art History and Head of Department (Art History & Art Conservation) at Queen's University, Kingston, Canada.

Beverly Lemire is Professor & Henry Marshall Tory Chair, Department of History & Classics, University of Alberta, Canada.

Alena Buis is Doctor of Philosophy Candidate in Art History, Queen's University, Kingston, Canada.

The Histories of Material Culture and Collecting, 1700–1950

Michael Yonan, University of Missouri-Columbia, USA

The Histories of Material Culture and Collecting provides a forum for the broad study of object acquisition and collecting practices in their global dimensions from 1700 to 1950. The series seeks to illuminate the intersections between material culture studies, art history, and the history of collecting. HMCC takes as its starting point the idea that objects both contributed to the formation of knowledge in the past and likewise contribute to our understanding of the past today. The human relationship to objects has proven a rich field of scholarly inquiry, with much recent scholarship either anthropological or sociological rather than art historical in perspective. Underpinning this series is the idea that the physical nature of objects contributes substantially to their social meanings, and therefore that the visual, tactile, and sensual dimensions of objects are critical to their interpretation. HMCC therefore seeks to bridge anthropology and art history, sociology and aesthetics. It encompasses the following areas of concern:

1. Material culture in its broadest dimension, including the high arts of painting and sculpture, the decorative arts (furniture, ceramics, metalwork, etc.), and everyday objects of all kinds.

2. Collecting practices, be they institutionalized activities associated with museums, governmental authorities, and religious entities, or collecting done by individuals and social groups.

3. The role of objects in defining self, community, and difference in an increasingly international and globalized world, with cross-cultural exchange and travel the central modes of object transfer.

4. Objects as constitutive of historical narratives, be they devised by historical figures seeking to understand their past or in the form of modern scholarly narratives.

The series publishes interdisciplinary and comparative research on objects that addresses one or more of these perspectives and includes monographs, thematic studies, and edited volumes of essays.

Craft, Community and the Material Culture of Place and Politics, 19th-20th Century

Edited by
Janice Helland, Beverly Lemire
and Alena Buis

ASHGATE

Published by

Ashgate Publishing Limited
Wey Court East
Union Road
Farnham
Surrey, GU9 7PT
England

Ashgate Publishing Company
110 Cherry Street
Suite 3-1
Burlington, VT 05401-3818
USA

www.ashgate.com

British Library Cataloguing in Publication Data
A catalogue record for this book is available from the British Library.

Library of Congress Cataloging-in-Publication Data
Craft, community and the material culture of place and politics, 19th–20th century / edited by Janice Helland, Beverly Lemire, Alena Buis.
pages cm. -- (The histories of material culture and collecting, 1700–1950)
"This volume emerged from a 2011 conference addressing the interconnections of material culture, craft and community, an event held at the University of Alberta and sponsored by the Material Culture Institute."
Includes bibliographical references and index.
ISBN 978-1-4094-6207-1 (hbk) 1. Material culture--History--19th century. 2. Material culture--History--20th century. 3. Technology--Social aspects. I. Helland, Janice, editor of compilation. II. Lemire, Beverly, 1950- editor of compilation. III. Buis, Alena.
GN406.C72 2014
306.09'034--dc23

2013031412

ISBN 9781409462071

MIX
Paper from responsible sources
FSC® C013985

Printed in the United Kingdom by Henry Ling Limited,
at the Dorset Press, Dorchester, DT1 1HD

Contents

List of Illustrations

2.4 Moosehair-embroidered hide moccasins, late 1840s. © Bata Shoe Museum, Toronto, Ontario. P82-0054-A.

2.5 Henry Daniel Thielcke, *The Presentation of a Newly-Elected Chief of the Huron Tribe, Canada*, 1841. © McCord Museum, Montreal, Quebec. M20009.

3 'To foster and encourage the study and practice of Māori arts and crafts': Indigenous Material Culture, Colonial Arts and Crafts and New Zealand Museums

3.1 Royal Doulton plate with Māori designs showing arts and crafts influences. The Māori word '*Kiaora*' means hello. Bone-china cabinet plate with printed decoration and hand-painted detailing, c. 1925. Overall dimension (height) 25 mm. Museum of New Zealand Te Papa Tongarewa GH011281.

3.2 *Kōwhaiwhai* pattern copied from rafters in a meeting house near Gisborne by Herbert Williams in the 1890s and reproduced in Augustus Hamilton's book *Māori Art* (1901). Herbert William Williams, 'Designs of ornamentation on Māori rafters No. 4, 5 and 6'. Three red and black ink drawings, each 82 × 350 mm. Alexander Turnbull Library E-331-f-004/006.

3.3 Photograph of the Native Affairs Committee in the Māori Affairs Committee Room, Parliament, Wellington 1922. Alexander Turnbull Library, Wellington, New Zealand. Ref: PAColl-8205.

3.4 The carvers from the School of Māori Arts and Crafts at work at the Dominion Museum, Wellington, 1936. William Hall Raine/Museum of New Zealand Te Papa Tongarewa B.13045.

3.5 The interior of the Auckland War Memorial Museum built in 1929, showing the Māori court at the centre of the building with major carved artifacts: a *pātaka* or storehouse (left), *wharenui* or house (centre) and *waka* or canoe (right). Photographer: Elmer R. Nelson. Original print received from Tāmaki Paenga Hira (id71951.tif). Courtesy of the Milwaukee Public Museum.

4 Reconstructing the Old Dutch House: New Castle's Colonial Dutch Gem and Louise Crowninshield's Preservation Community

4.1 The Dutch House, New Castle, Delaware, 21 September 2012. Courtesy of the New Castle Historical Society.

4.2 The Dutch House, New Castle, Delaware, undated postcard. Courtesy of the New Castle Historical Society.

4.3 Photograph of H.F. du Pont and Louise Crowninshield. Courtesy of the Winterthur Library: Winterthur Archives.

4.4 The Dutch House, Interior, 2012. Courtesy of the New Castle Historical Society.

5 From the Gift Shop to the Permanent Collection: Women and the Circulation of Inuit Art

5.1 'Women's Handicraft Guild members Mrs. J. Strachan (left) and Mrs. J. Harley display Eskimo handiworks', from the Edmonton *Bulletin*, November 25, 1950. City of Edmonton Archive EA-600-6204.

5.2 Photograph of Mrs Robert Miller, from the *Life* magazine article 'Church Sale of Arctic Art', May 25, 1959, p. 43. Photograph: Fritz Goro/Time & Life Pictures/Getty Images.

5.3 The Gift Shop at the Winnipeg Art Gallery, 1979. Photograph taken by David Barbour. Courtesy of the Winnipeg Art Gallery.

5.4 Victoria Mamnguksualuk, *Keeveok's Journey*, 1969, Stencil on paper, Collection of the Art Gallery of Alberta, purchased with funds from the Women's Society of the Edmonton Art Gallery.

Notes on Contributors

Alena Buis is a PhD Candidate in the Art Department at Queen's University in Kingston, Ontario. She completed a MA degree in Canadian Art History at Concordia University in 2008. Her current dissertation research, funded by the Social Sciences and Humanities Research Council of Canada, focuses on seventeenth-century Dutch material culture in the North American colonies. Buis's publications include a chapter entitled 'The Raw Materials of Empire Building: Depicting Canada's "Home Children"', in *Depicting Canada's Children* (2008) and '"A Story of Struggle and Splendid Courage": Anne Savage's CBC Broadcasts of *The Development of Art in Canada*', in *Rethinking Professionalism: Women and Art in Canada: 1850–1970.* She has also contributed articles to *Inuit Art Quarterly, de Halve Maen, Dutch Crossing, Cahiers métiers d'art* and *Material Culture Review.*

Conal McCarthy is Director of the Museum & Heritage Studies programme at Victoria University of Wellington, New Zealand. He has degrees in English, Art History, Māori language and Museum Studies, and has worked in galleries and museums in a variety of professional roles as well as sitting on the boards and advisory groups of a number of cultural institutions. He has published widely on Māori art, museum display and current museum practice, including the monograph *Museums and Māori: Heritage Professionals, Indigenous Collections, Current Practice* (Left Coast Press, 2011). His next book is *Museum Practice: Critical Debates in Contemporary Museums* (Wiley-Blackwell, 2014).

Edward S. Cooke, Jr, the Charles F. Montgomery Professor of American Decorative Arts in the Department of the History of Art at Yale University, has published extensively on historical and contemporary craft and material culture. He is also one of the founding editors of *The Journal of Modern Craft.*

Lily Crowther is a curator specializing in modern craft and design. She was educated at Magdalen College, Oxford, and Essex University. Since 2006, she has worked at the Victoria & Albert Museum, London, on projects including new permanent galleries of twentieth-century design and studio ceramics; most recently, she was involved in the exhibition 'British Design 1948–2012', staged to mark the occasion of the London 2012 Olympics. She has published on mid-century modern British design, and on Coventry Cathedral, as well as on the links between craft practice and suburbia.

Annette de Stecher has a PhD in North American Indigenous Art History from the Institute for Comparative Studies in Literature, Art and Culture at Carleton University. Her dissertation explored the visual arts of the Wendat of Wendake, Quebec, focusing on their nineteenth-century souvenir arts tradition. Her current research broadens this study to explore regional visual arts traditions in the Eastern Woodlands. In 2011 de Stecher received a Research Fellowship at the Canadian Museum of Civilization; in 2012, as Curator of Inuit Art at Carleton University Art Gallery, she curated the exhibition, *The Past is Present: Memory and Continuity in the Tyler/Brooks Collection of Inuit Art*, which opened in June 2013.

Janice Helland is Professor (Art History) and Head of Department (Art History and Art Conservation), Queen's University, Canada. Her publications on British and Irish women artist/designers and craft organizations include *British and Irish Home Arts and Industries 1880–1914: Marketing Craft, Making Fashion* (Dublin, 2007) as well as articles and book chapters. Together with Sandra Alfoldy she edited *Craft, Space and Interior Design, 1855–2005* (2008) and, with Bridget Elliott, *Women Artists and the Decorative Arts, 1880–1935: The Gender of Ornament* (2002).

Beverly Lemire is a Fellow of the Royal Society of Canada and Professor of History and Henry Marshall Tory Chair in the Department of History & Classics and the Department of Human Ecology, University of Alberta. She is founding Director of the Material Culture Institute at that university. Her publications include *Fashion's Favourite: The Cotton Trade and the Consumer in Britain, 1660–1800* (1991), *Dress, Culture and Commerce: the English Clothing Trade before the Factory* (1997), *The Business of Everyday Life: Gender, Practice & Social Politics in Britain 1600–1900* (2005, 2012 pb) and *Cotton* (2011) in the series Textiles that Changed the World. Edited works include *Women and Credit: Researching the Past, Refiguring the Future* (2001) and *The Force of Fashion in Politics and Society: Global Perspectives from Early Modern to Contemporary Times* (2010).

Alla Myzelev is an adjunct professor at OCAD University and University of Guelph. She is currently working on a single-authored manuscript, *Creating Modern Living in Toronto: From Vernacular to Deco*, and she is one of the contributors to the Nelson's Modules project on Canadian History. With John Potvin she edited *Material Cultures, 1740–1920: The Meanings and Pleasures of Collecting* (2009) and *Furniture, Interior Design, and Contours of Modern Identity* (2010). She has published numerous articles and essays on Russian and Ukrainian avant-garde, British and Canadian design and architecture, and the knitting revival in, for example, *Craft Research*, *Textile: Journal of Cloth and Culture*, *Journal of Modern Craft* and *Craft Journal*. Myzelev is a recipient of several prestigious grants, among them an SSHRC post-doctoral fellowship (2007) and a Paul Mellon Research Fellowship (2008).

Vivienne Richmond is a lecturer in Modern British History at Goldsmiths, University of London. She is the author of *Clothing the Poor in Nineteenth-century England* (Cambridge University Press, 2013), several related articles and chapters, and co-editor (with Clare Rose) of *Clothing, Society and Culture in Nineteenth-century England* (Pickering & Chatto, 2011).

Anne Whitelaw is Associate Professor of Canadian Art in the Department of Art History at Concordia University in Montreal. Whitelaw has published extensively on the display of Canadian art at the National Gallery of Canada, on the integration of aboriginal art in national museums, and on Canadian art historiography. She is co-editor with Brian Foss and Sandra Paikowsky of *The Visual Arts in Canada: The Twentieth Century* (2010). Whitelaw's current research includes a book on the relationship between federal cultural institutions and art galleries in Western Canada, and an exploration of the work of women's volunteer committees in North American museums.

Claire Wintle is a senior lecturer in the History of Art and Design at the University of Brighton, UK. She has published in journals such as *Victorian Studies* and *History Workshop Journal*, and her book, *Colonial Collecting and Display: Encounters with Material Culture in the Andaman and Nicobar Islands* (2013), is published by Berghahn. Wintle's research is increasingly focused on museums and decolonization, 1945–70, and Indian cultural diplomacy in the twentieth century.

Acknowledgements

This volume emerged from a 2011 conference addressing the interconnections of material culture, craft and community, an event held at the University of Alberta and sponsored by the Material Culture Institute. The Social Science and Humanities Research Council of Canada provided generous funding. The University of Alberta also delivered timely support of many kinds, in addition to that given by the Material Culture Institute. The event was realized through the collective energies of the organizing committee headed by Beverly Lemire, along with Catherine Cole, Lianne McTavish, Arlene Oak, Megan Strickfaden and Sarah Waurechen. Melanie Marvin supplied unstinting and invaluable administrative assistance at every turn. Alena Buis, Janice Helland and Beverly Lemire built on the foundation of this gathering to create this volume. It was a great pleasure to work with editor Margaret Michniewicz, of Ashgate, who smoothed the process of production. Julie Hollenbach, Janice Helland's research assistant, provided assistance at every stage of the editing process. Her commitment and meticulousness were invaluable, while her humour and calm helped keep everything on track. Finally, Kristine Kowalchuk stepped in to handle the index with her usual aplomb and precision. She has earned our thanks on this and other occasions.

Every attempt has been made to obtain permission to reproduce copyright material. If any proper acknowledgement has not been made, copyright holders are invited to inform the editors of the oversight.

Introduction

Janice Helland, Beverly Lemire and Alena Buis

Material culture study is now a burgeoning interdisciplinary field, enabling exceptional insights through the critical study of objects. This field flourishes through a vibrant cross-pollination of methodologies and theoretical perspectives from various disciplinary foundations. The study of craft production and craft objects fits within the wider heterogeneity of material culture study, approached with a range of interdisciplinary questions. More recently 'craft' has become a subject of sustained focus in history, art history, design studies, anthropology, art practice and material culture studies, generating a rich diversity of publications, events and exhibitions. With object study at the core, this collection addresses the past and present of craft production, use and meaning within a range of community settings. Place matters, as does the time period and institutional or community context of object production, use, display or transfer. These factors engender different meanings when assessing handcrafted, as opposed to industrially produced, materials. All of these elements figure in the discussions of craft in this volume. Craft practice has a rich history and remains vibrant, sustaining communities while negotiating cultures within local or international contexts. The more than two centuries of industrialization have not extinguished handmade goods. However, the broader force of industrialization inevitably redefined and continues to define the context of creation, deployment and use of craft objects. Craft production has flourished and continues to do so despite the powerful juggernaut of global industrialization, whether inspired by a calculated refutation of industrial sameness, an essential means to sustain a cultural community under threat, or a refusal of imposed definitions by a dominant culture. The broader effects of urbanizing, imperial and globalizing projects shape the multiple contexts of interaction and resistance that can define craft ventures through place and time.

This publication arises from an earlier collective project, the 2011 conference 'Material Culture, Craft & Community: Negotiating Objects Across Time & Place'. This event, sponsored by the Material Culture Institute at the University of Alberta, showcased the wide range of research addressing craft and community. The conference attracted speakers from almost every continent, from various professions and stages of career. They illustrated the academic attention now paid to craft and the important advancements of this field of study; and the conference provided a site for the presentation of research that considered the negotiation of crafts from a host of disciplinary perspectives. The conference programme allowed a comparative framework from which to discern unique and common features, to identify gender, ethnic and economic challenges, to chart the effects of globalization and the spaces and places within the market that enable craftmaking to survive and thrive. The conference also uncovered the interactions within and between cultures that come with commodity flows, as reflected in goods from furnishings to fine needlework, as well as the varied contexts of indigenous craftmaking and meaning. The common intellectual thread was an interest in the material culture of craft and its cultural and social disposition within societies. These scholars confirm the dynamic nature of this research area. The chapters offered here represent the rich developments within this field and point to important avenues of inquiry, where community values, dynamisms and priorities are expressed through craft. The varied social lives of craft objects are recorded and assessed in this volume, along with the peoples and institutions through whose hands they passed. The circumstances of craft production vary enormously; but withal a collective context is evident, a critical place from which to assess commonalities and differences.

As Edward Cooke notes in his essay, the term 'village', a word ripe with connotations, 'when linked with craft has a long tradition in modern discourse' dating back to the late nineteenth century, differentiating the context of production from the industrial. Like 'village', 'community' has similar, often romanticized, but frequently positive associations: cohesive groupings, cooperative labour, collective production. By addressing ways craft practices can reinforce, undermine and re-imagine cultural identities, the essays that follow seek to nuance idealizations of community as not stagnant, homogeneous groupings but rather dynamic, fluid groups of people and things, organizing and re-grouping at various times and places. Craft practice has long defined communities and groups, and continues to do so in the midst of global trade networks. In recent years, there has been great interest in the role of crafts within the concept of globalization, frequently utilizing village and community stereotypes to delineate so-called exotic traditional culture that made curious objects from more modern cultures and their informed consumers. However, rather than engage in binary distinctions of western/non-western, we are interested in addressing examples of local craft production that cross boundaries because of colonialism or imperialism, with craft production in a globalized world (either contemporary or historical) but always with a focus on the relationship between craft and the community in which it was produced. As such, case studies of craftmaking in contact zones highlight issues such as oppression and subjugation, power and domination, allowing analyses to be complicated and expanded. To quote Lynette Russell's study of colonial frontiers: 'Cross-cultural encounters produce boundaries and frontiers. These are spaces, both physical and intellectual, which are never neutrally positioned, but are assertive, contested and dialogic'.[1]

Craft as material culture often has been and frequently is still produced in communities and, as such, participates in the construction of a community identity. One needs only to view a Navajo weaving, for example, to recall that the object is associated with a specific location and a specific group of people who live and work in the southwest part of the United States. The object might be historical and on view in a gallery, or it could be contemporary and for sale at a tourist destination; either way the object speaks of a relationship between place and producer. Additionally, and as with Navajo weavings, these relationships may be contested and political. Many weavings that tourists classify as Navajo are now made in Mexico and can be sold for considerably less than an 'authentic' weaving.[2] In this case the Navajo community may work toward educating consumers about the production of the weaving or introduce a copyrighted trademark so that the objects can be identified. Historically, a similar trajectory can be traced with the weaving of tweed in the Scottish Highlands when a British public was taught to recognize hand-woven tweed by the quality of the weave and the trademarked cloth as different from mass-produced tweed cloth.[3] Again, the trademark identified a specific community of weavers and the introduction of the trademark signalled a political act meant to channel profits back to the weavers themselves and persuade consumers to look for quality and authenticity, both of which are contested terms but which also can indicate a self-aware and self-directed form of collective campaign.[4]

The making of craft, particularly when associated with a specific community, can emerge as political intent. According to Annette Weiner and Jane Schneider in their introduction to *Cloth and the Human Experience* (1989), craftmaking can be political and even subversive, 'above all in societies emerging from colonial domination'.[5] The chapters in this book address the politics of craft and colonization but additionally consider how craft production can be seen as political within communities defined by various configurations of class and gender. Craft as material culture can speak of community in geographic locations as diverse as urban London suburbs and rural Wendat (Huron) villages or islands in the Bay of Bengal and trading outposts in colonized North America. Our objective is to explore relationships between craft and community as related to place and politics without privileging a specific chronology or location. Thus, while a number of the chapters discuss material culture made in communities during the nineteenth to early twentieth century and exhibit characteristics that retrospectively may be considered part of the Arts and Crafts movement,[6] the chapters are not presented in strict chronological order; rather we sought to discover themes and relationships that occurred without invoking the frequently understood 'progressiveness' of the temporal. This is particularly important when investigating diverse cultures and societies. Additionally, and for similar reasons, we avoided establishing groupings according to geographical locations and, furthermore, the essays address communities that, while geographically specific, were not restricted to a specific place. Therefore the book opens with Edward Cooke's chapter, which elaborates on rural craft and village craft within a global economy by focusing on the exhibitions of Indian and Japanese Villages that took place during the 1880s, through to the material culture of the Southern Highlands in the United States in the early part of the twentieth century. And we end with Alla Myzelev's chapter – a discussion far removed from the rural and the village – which examines a planned and organized place in the suburban environment of a rapidly expanding metropole as a craft collective that facilitated dialogues of community and community values. A question posed by all the studies presented here is whether changes or alterations in the relationship between craft and community arose organically within the community itself or operated as a response to external influences and pressures that shaped priorities into material forms. As Edward Cooke suggests in his chapter, the history of craft and community is complicated; in exploring this complication, he observes both the 'potentially positive and sinister aspects of globalized craft'.

Craftmaking as community production sometimes emanates from inside the community, as it did with the Wendat peoples of Quebec during the difficult economic transition at the end of the fur trade era, as incoming European migrants drastically constrained their access to hunting lands. The Wendat, strategically and with skill, developed hybrid designs that incorporated their own traditions while at the same time appealing to settlers and tourists. As Anne de Stecher observes, Wendat souvenir craft played a role in sustaining community and identity, thereby 'demonstrating the continuity of Wendat traditions as they responded with forethought and strategic planning to the increasing pressures of colonizing settler presence'. As early as the seventeenth century the political acumen of the community, the traditions of diplomacy and the understanding of the importance of trade laid the basis for what became an internationally successful and financially viable craft production by the mid-nineteenth century. The 'craft-art' made by Wendat artisans spoke of community but also of the vibrant role that objects played in defining a culture and an aesthetic; the objects themselves can be seen to represent a political and economic strategy deployed in a contact zone to sustain a community and its identity as political circumstances evolved.

Māori people in New Zealand, somewhat like the Wendat in Canada, assumed a role in the changes and participated in the transformation of the production of handcrafted objects. As Conal McCarthy demonstrates in his chapter on the material culture of Māori, the transformation of Māori objects from ethnographic categorization as curios or specimens made by colonized 'savages' to their introduction into museums as Māori arts and crafts emanated from both within and without the community itself. McCarthy argues that 'Māori "art" was not *discovered* but *produced*, and Māori were involved in this production because it allowed them to engage with settler society'. Indigenous politicians such as Āpirana Ngata (1874–1950) took advantage of the government-funded Board of Māori Arts to maintain and develop craft production as a negotiation between indigenous traditions and western art practices. During the 1920s and 1930s the School of Māori Arts and Crafts became inextricably linked with the national museum and, thus, McCarthy contends that the craft objects functioned to promote tribal identity even as they served government assimilationist policies. Carving and weaving became objects that signalled cultural pride and celebrated survival within the contested site of colonization.

The European diaspora also provided examples of communities utilizing material culture from within to retain the familiar elements of home while they interacted with an already highly developed indigenous culture and later with another wave of colonizers. Accordingly, Alena Buis in her chapter focuses on the gathering of early colonial material culture used to furnish Delaware's historic Dutch House originally constructed c. 1690–1710 to investigate the gendered communities formed by the early twentieth-century Colonial Revival movement. Although the revisionist movement and the connected Holland Mania created a highly problematic sanitized view of early

American life, through exhibitions, publications and the establishment of historic sites, it did bring a sustained interest in material culture, interiors and the preservation of historic crafts. Led by wealthy philanthropist Louise Crowninshield, but involving many people with ties to the Colonial Revival movement, Buis analyses the collaborative social construction of the Dutch House as an example of the staged and imagined symbolic community identities created through nostalgic views of early American colonization. Her chapter describes the ways the small house museum contains two intertwined narratives: primarily the town's early Dutch history, but secondarily also the ways collectors and preservationists of the early twentieth century interpreted early colonial life. As a critical investigation of Crowninshield's craft collections and her involvement in projects such as the Dutch House, this case study articulates 'the gendered politics of early twentieth-century professionalization of cultural institutions'.

In other delineated places, craft can generate a layered and complicated identity that has its impetus imposed upon a community from without, as in the case of middle- and upper-class women promoting the acquisition and sale of Inuit art in Canada. In her chapter on museum gift shops Anne Whitelaw suggests that female members of voluntary museum organizations have played a major role in the circulation and acquisition of Inuit craft since its introduction to the North American market. Whitelaw provides a case study of how a colonizing culture assumes the role of patron and thereby becomes instrumental in directing the transformation of local craft to global commodity. Inuit art was first offered for sale in the 1950s in gallery gift shops operated by volunteers, then frequently purchased by volunteers and given as donations to institutional permanent collections. Whitelaw argues that the relatively powerless and gendered space occupied by museum volunteers gave women a sense of agency and purpose when they promoted specific cultural practices. Although their activities were closely 'tied to the aims of the larger institutions they supported, in the realm of Inuit art (or Eskimo handicraft as it was then described) women played active roles in deciding how the work would be marketed and sold, and were instrumental in arguing for the legitimacy of the inclusion of such objects in the collections of art galleries'.

Another manifestation of identity imposed upon a locale or community can be found among organizations that prided themselves on either establishing a crafting community or providing specific groups of artisans with opportunities to market their products. Late nineteenth-century Britain spawned a number of so-called philanthropic organizations that aimed to direct the activities of artisans to produce objects that would be attractive to sophisticated urban buyers while, at the same time, eliminating middlemen and returning profits to craft workers. Janice Helland discusses Londoner Alice Hart, who established the Donegal Industrial Fund in 1883 to assist craftspeople in remote areas of Donegal in Ireland. Hart successfully transported handcrafted objects (many of which were textiles) from Donegal to London, where she exhibited and sold the goods without profit; she quickly expanded her market to include the United States. Although the Fund is usually considered a philanthropic organization, Hart's objective was to establish what today would be called a 'fair trade network' that would benefit artisans otherwise at the mercy of middlemen. This chapter discusses Alice Hart's Donegal Industrial Fund as a community of artisans who participated in an historical 'fair trade network' and in so doing suggests that although 'fair trade' is usually regarded as a post-WWII phenomenon, traces of such networks can be found much earlier. Additionally, Hart consistently promoted the objects the Fund sold as typically Irish, emphasizing a national character in these goods, terming the people who produced them as 'strikingly intelligent' but misunderstood by the English. She related the making and selling of the works to the 'political question' and the 'economic question' while, at the same time, she acknowledged the community: 'In this undertaking success would have been impossible, had not the people themselves ably and admirably seconded every effort made'.[7]

Claire Wintle's chapter on the making by the Great Andamanese people of craft objects that were often characterized as 'fakes' also addresses political and economic questions while, at the same time, she introduces a narrative of community production, subversion and 'an everyday form of resistance'. Wintle suggests that despite external influences wrought by the colonial encounter, the crafted objects 'were firmly rooted in Andamanese design histories and worldviews'. Andamanese material culture may have been produced for personal use or for trade, but the process of making was 'socially learnt and culturally embedded'. Therefore, despite the location of production in resident homes established by the British, Andamanese crafting spoke of a cultural cohesiveness that was maintained in objects despite colonial intervention. Furthermore, within the restrictive environment new stylistic trends emerged that should not be labelled 'inauthentic' but that rather, as Wintle suggests, constitute an 'instance of Andamanese vitality and fluidity'.

While the Andamanese were placed into resettlement homes that were considered civilizing institutions, invalided women in England were confined to their homes as sites of production because they were unable to complete in the marketplace with the fully abled. The Girls' Friendly Society, founded in London in the late nineteenth century, was an organization that constructed a community for young working-class women who could produce objects, mostly embroidery, in their own homes under the auspices of the Society, which, in turn, provided a depôt where works could be sold. As Vivienne Richmond establishes in her chapter, the Society began operating in 1875 as an association that catered for all chaste working-class women but quickly established a special department for members with chronic illnesses and disabilities. By the late nineteenth to early twentieth century, this department, which also operated the Central Needlework Depôt, offered a selection of correspondence courses: it 'taught "arts and crafts", particularly sewing, to "invalid" members, then employed them as home-workers to produce goods, especially clothing, for sale at its London showroom'. This community did not occupy a specific geographic location, but rather the women were united by their membership in and connection to the Depôt. Richmond convincingly argues that instead of 'marking out' its invalid workers, the Depôt served a 'normalising function by allowing them to participate in the same productive labour as many of their able-bodied sisters'. It took account of the workers' limitations and accommodated their health problems, thereby providing them with what in today's terms might be characterized as a 'safe space'. Richmond does not romanticize the organization or deny the inequalities evident between organizer and worker; rather she initiates discussions into an under-researched area of study and does so by interrogating the community of workers associated with the Depôt and the work they produced.

Lily Crowther's chapter introduces a discussion of larger built environments, specifically urban communities, which command an identity that develops from within the space of a suburb or neighbourhood. Here artisans and the building trades contributed to place and space as they constructed or renovated local areas. Crowther explores urban areas around arts schools such as the Camberwell School of Arts and Crafts located in south east London during the late nineteenth to early twentieth century as repositories and places of training for craftspeople. She investigates how Camberwell and other similar schools provided a link between the craft of building and the community, and argues that buildings can also be considered collective craftworks. The building of planned suburbs by trained craft workers might be viewed as neighbourhoods that speak of collaborative craftwork and enable discussions of community values. Crowther moves our thinking about craft and place into the suburban environment, into the realm of built objects that form parts of the city and its material spaces and, in so doing, asks us to dramatically complicate the definition of both craft and material object. Does the craftwork in an urban space provide a unifying and identifiable cultural space with which inhabitants identify, and how is this made? Indeed, asks Crowther, 'how is a city made?'

The book closes with Alla Myzelev's discussion of an independent community that, as in Crowther's study, is suburban with close ties to the metropole. She explores an organized space made specifically for the production of craft in Canada during the Great Depression of the 1930s – the Guild of All Arts established in 1932 in a Toronto suburb as a craft community with workshops open to the public and artists living on the premises. It included a profitable hotel that offered some of its fixtures for sale, the headquarters of an innovative weaving method using portable looms and, finally, a depository of architectural remains from the Toronto building boom of the 1920s and 1930s. Thus Myzelev investigates relationships between communal efforts to practice craft and attempts to make the enterprise financially viable. Many craftspeople and artists who were struggling during the Depression came to live in Guildwood, as it became known, and produced objects that were later sold in craft outlets. The objects included textiles, woodcarvings, ceramics and stone sculptures, all of which were meant to showcase 'Canadian material and local flora, fauna and history'. While this chapter is concerned with an economic reality more than a political space, its members, although struggling financially, were, nevertheless, part of a privileged settler society, and their attempts to establish a 'Canadian' style were exclusionary. Today, as Myzelev suggests, the location that was the Guild of All Arts has become 'a cemetery of forgotten buildings' and the community of craft producers can be found only in 'traces of workshops and architectural remains'.

Conclusion

The concepts of the 'artisan', 'craft' products and 'handmade' goods are once again a matter of public discussion outside the academy, as even in developed nations young men and women turn their hands to the making of craft commodities both for the qualities imbued in the goods themselves and for the context of their production. *The New York Times* reported recently that 'Contrary to popular belief, the revival of craft manufacturing [in the United States] isn't just a fad'.[8] In the hyper-competitive contemporary global marketplace, specialist craft production is now recognized as an effective strategy to sustain individuals, groups and communities, whether recycling juice boxes into fashionable bags as in a Philippines cooperative project, making craft foods in Brooklyn for local aficionados or practicing the mysteries of pre-industrial paper production.[9] The power of handcrafted objects rests not only in their aesthetic or functional features but also in the political and cultural values embedded in these things. As all the chapters in this book suggest, attending to the political histories of these goods and their makers reveals the creative persistence of various hand mediums and the material debates they represented.

Sources Cited

Davidson, Adam, 'It Ain't Just Pickles: Craft Business May Portend the Future of the U.S. Economy', *The New York Times Magazine* (19 February 2012), p. 14.

Hart, Alice, *The Cottage Industries of Ireland, with an Account of the Work of the Donegal Industrial Fund* (London: Hamilton, Adams & Co., 1887).

Hart, Imogen, *Arts and Crafts Objects* (Manchester: Manchester University Press, 2010).

Helland, Janice, *British and Irish Home Arts and Industries 1880–1914: Marketing Craft, Making Fashion* (Dublin: Irish Academic Press, 2007).

Levine, Mark, 'Cellulose Hero', *The New York Times Magazine* (19 February 2012), pp. 26–9.

McLerran, Jennifer, *Weaving is Life: Navajo Weavings from the Edwin L. & Ruth E. Kennedy Southwest Native American Collection* (Seattle, WA: Kennedy Museum of Art, Ohio University, 2006).

M'Closkey, Kathy, 'Part Time for Pin Money', in Kimberly M. Grimes and B. Lynne Milgram (eds), *Artisans and Cooperatives: Developing Alternative Trade for the Global Economy* (Tucson, AZ: University of Arizona Press, 2000), pp. 143–58.

Milgram, B. Lynne, 'Reclaiming Materials and Fashion-work in the Urban Philippines', in Beverly Lemire (ed.), *The Force of Fashion in Politics and Society: Global Perspectives from Early Modern to Contemporary Times* (Aldershot: Ashgate, 2010), pp. 191–214.

Russell, Lynette (ed.), *Colonial Frontiers: Indigenous–European Encounters in Settler Societies* (Manchester: Manchester University Press, 2001).

Stevens, Dennis, 'Validity Is in the Eye of the Beholder: Mapping Craft Communities of Practice', in Maria Elena Buszek (ed.), *Extra/Ordinary: Craft and Contemporary Art* (Durham, NC and London: Duke University Press, 2011), pp. 41–58.

Weiner, Annette and Jane Schneider (eds), *Cloth and the Human Experience* (Washington, DC: Smithsonian Books, 1989).

Notes

1 Lynette Russell (ed.), 'Introduction', *Colonial Frontiers: Indigenous–European Encounters in Settler Societies* (Manchester University Press, 2001), p. 1.

2 Jennifer McLerran, *Weaving is Life: Navajo Weavings from the Edwin L. & Ruth E. Kennedy Southwest Native American Collection* (Kennedy Museum of Art, Ohio University, 2006); Kathy M'Closkey, 'Part Time for Pin Money', in Kimberly M. Grimes and B. Lynne Milgram (eds), *Artisans and Cooperatives: Developing Alternative Trade for the Global Economy* (University of Arizona Press, 2000), pp. 143–58.

3 Janice Helland, *British and Irish Home Arts and Industries 1880–1914: Marketing Craft, Making Fashion* (Dublin: Irish Academic Press, 2007).

4 Although essays in this book do not deal with contemporary craft practices, it is worth noting that 'craft communities' remain vital if somewhat reconfigured. Dennis Stevens, for example, in his investigation of craft production, writes that 'craft communities today, although tradition is acknowledged ... are not adhering to conventional expectations. A new variable has entered the domains of practice: the rise of citizen journalism and what the media has coined the "Netroots" movement'. D. Stevens, 'Validity Is in the Eye of the Beholder: Mapping Craft Communities of Practice', in Maria Elena Buszek (ed.), *Extra/Ordinary: Craft and Contemporary Art* (Durham, NC and London: Duke University Press, 2011), p. 47.

5 Annette Weiner and Jane Schneider (eds), *Cloth and the Human Experience* (Washington, DC: Smithsonian Books, 1989), p. 4.

6 Imogen Hart in her study *Arts and Crafts Objects* (Manchester University Press, 2010) interrogates the concept 'Arts and Crafts movement' and pushes the beginnings of the interest in art, craft and design usually associated with the 'movement' to the 1830s in the UK.

7 Alice Hart, *The Cottage Industries of Ireland, with an Account of the Work of the Donegal Industrial Fund* (London: Hamilton, Adams & Co., 1887), p. 9.

8 Adam Davidson, 'It Ain't Just Pickles: Craft Business May Portend the Future of the U.S. Economy', *The New York Times Magazine*, 19 February 2012, p. 14.

9 B. Lynne Milgram, 'Reclaiming Materials and Fashion-work in the Urban Philippines', in Beverly Lemire (ed.), *The Force of Fashion in Politics and Society: Global Perspectives from Early Modern to Contemporary Times* (Aldershot: Ashgate, 2010), pp. 191–214; Davidson, 'It Ain't Just Pickles', pp. 14–17; Mark Levine, 'Cellulose Hero', *The New York Times Magazine*, 19 February 2012, pp. 26–9.

1

Village Craft, Rural Industry: The Politics of Modern Globalized Craft

Edward S. Cooke, Jr

Over the past decade there has been great interest in the role of crafts and craftwork within the concept of globalization. As Tom Friedman and other columnists have observed, we live in a flat economic world in which an interpenetrating cultural landscape has been enabled by technology and Globalization 3.0. Many craft organizations have seized upon the possibilities of economic growth implicit in the world of multinational companies and commodity chains in which a t-shirt is made from cotton grown in the United States, spun and woven in China, customized in the Caribbean, and sold in the United States, only to then be recycled as jansu cloth used for bedding in Nigeria.[1] Within this global framework, craftwork occupies a niche as a place-specific commodity that is produced in a specific location, often outside of urban areas, by an identifiable maker or group of makers in a small shop reliant on dexterity rather than expensive specialized machinery. Key to the creation and sustenance of this market is a reliance on rhetoric and imagery to reinforce the distinction between the authentic object produced by an idealized producer and the placeless goods of impersonal industrial capitalism.

Taking advantage of the information revolution, many craft businesses have sought to use the internet to link isolated indigenous artisans in Africa, South Asia, Oceania and South America with markets in more prosperous areas. However, such economic optimism has also precipitated a critical response. Some craft leaders have characterized globalization as a form of neocolonialism in which interdependence is lop-sided or asymmetrical. Others question whether artisanal 'outsourcing' simply commodifies the craft product with no consideration of the cultural form and content of its origins. One might refer to this conundrum of modern global craft simply as the question of fair trade for creative labour. Instead of economically or ecologically sustainable development, these critics argue for a comprehensive cultural sustainability to improve the beneficial reciprocity of globally based craftwork. Such an endeavour begins with attention to the social conditions of the makers and the national protection of self-reliance and self-sustainment. It also has to recognize that tradition is in fact dynamic. As Kevin Murray reminds us, 'cultural exogamy' can be a positive force.[2]

The tension between an external entrepreneurial and commercial spirit and an internal cultural inclination, that is between a colonial perspective and a post-colonial one, can be clearly seen in one company with a long history in marketing global handcraft – Ten Thousand Villages. With a mission to provide 'vital, fair income to Third World people by marketing their handicrafts and telling their stories in North America', the enterprise promises fair remuneration while preserving the narrative of the authentic craftsman.[3] Most telling about the organization's intent is the use of the noun 'village' in its name. The term implies a traditional form of social organization and the associated desirable standards of workmanship. 'Village' thus suggests authenticity and provides understandable criteria to assess the value of the objects produced in that environment. The example of Ten Thousand Villages seems an ideal catalyst to historicize and theorize the notion of village as the location of craftwork and gain a longer view of modern craft marketing.[4]

The Indian Village

When linked with craft, the word 'village' has a long tradition in modern discourse dating back to the 1880s, a period when British manufacturers and imperial officials rewrote the history of the production of household goods. In the seventeenth and early eighteenth centuries, the decentralized manufacturing of India, China and Japan in particular, relying on a large labour force intensively and skilfully using established technologies, had dominated the world markets in textiles and ceramics. Threatened by the massive capacity of the South and East Asian systems, Britain took a different direction dependent upon mechanization, industrial organization and capitalization to compete in the world markets. At the same time it created a false history of production that marginalized the Asian artisans as inefficient and non-productive in comparison with the efficiency and output of industrial production. Indian weavers and block printers and Chinese potters were not given a particular place in the concept of epochs of production popular at the time, but were pigeonholed as 'timeless' in an ahistorical sense. Expunging the capabilities of the Asian craft system thus permitted the creation of a celebratory history of Western technology, part of what Prasannan Parthasarathi includes in the concept of 'the great divergence'.[5] To borrow a metaphor from the days of vinyl 45s, village craft was the oft-forgotten dreamy flip side of British industry, which was the hit single.

The contrast between the technological superiority of British manufacturing machinery, so celebrated from the Crystal Palace Exhibition of 1851 onward, and the quaintness of timeless craft rooted in the concept of a village-based peasant system emerged in a public forum in the 1880s. As Saloni Mathur has argued regarding perception of Indian craft in the metropole during the decade, commercial interests fuelled a marginalized presentation of traditional Asian artisanal work. To market Indian-inspired goods, the London luxury goods retailer Liberty's established a 'living village of Indian artisans' in Battersea Park in November of 1885.[6] In developing this living display of craft, Liberty's commercialized George Birdwood's assertion that the village lay at the heart of the skill and beauty of Indian artefacts. Birdwood extolled the democratic values of village-based 'traditionary arts', in contrast to the fashionable values of the court-based 'sumptuary arts':

> To the Indian land and village system we altogether owe the hereditary cunning of the Hindu craftsman. It has created for him simple plenty, and a scheme of democratic life, in which all are coordinate parts of one undivided and indivisible whole, the provision and respect due to every man in it being enforced under the highest religious sanctions, and every calling perpetuated from father to son by those cardinal obligations on which the whole hierarchy of Hinduism hinges.[7]

For Birdwood, tradition and the mixed agricultural economy of the village thus freed the Indian artisan from 'the desperate struggle for existence which oppresses the life and crushes the very soul out of the English working man' in an industrial capitalist economy.[8] The Indian craftsman existed within a fuzzy ahistorical dimension, while the English craftsman worked within a very specific environment in regard to place and time.

While Birdwood might be considered a romantic preservationist, Arthur Liberty parlayed an interest in Indian craftsmen to develop the marketing scheme at Battersea Park (Figure 1.1). Housed in buildings that were purportedly 'accurate representations of Indian architecture', the craftsmen and entertainers in the Battersea village consisted of 32 inhabitants demonstrating 'typical Hindoo industries', including a sandalwood carver, dhurrie weaver, gold and silk embroiderer, potter who made 'beautiful objects of simple character' on his traditional kick wheel, snake charmer, knife tosser and nautch dancers, but the concept flopped due to the inadequate provision of salary, housing, food or clothing for the artisans and to a bitterly cold winter.

Regardless of the actual outcome, the emphasis seemed to be more on the romanticization of the concept of village craft than on the education of the consuming public, sales of the actual products, or situation of the craftsmen. Liberty's was trying to create the context for the sale of its own work, some of which was made in the UK and some in India (especially the Kutch region), and paid little attention to even fulfilling their promises to pay and care for the artisans. In this advertising campaign, otherness, village and handcraft were inextricably linked as spectacle.

The concept of demonstrating craftsmen was recycled, in altered form, the following year for the Colonial and Indian Exhibition. Organized by the leaders of the Indian Museum and the South Kensington Museum as a prelude to Queen Victoria's jubilee, the exhibition opened on May 4, 1886 and closed that November. The Indian Section consisted of a number of distinct displays, in order of space allocation, from largest to smallest: the Central Court, or Provincial Art-Wares Section, that displayed fine and decorative arts arranged according to different regions; an Imperial Economic Court that took a Natural History Museum angle with a selection of raw products from South Asia and clay figures of the various tribal people of British India; a space for commercial exhibitors of tea, coffee and tobacco; the Administrative Courts, which included a picture gallery with paintings and drawings; the Indian Vestibule; the Indian Palace, Courtyard and Shops; the Silk Court; and the Tea House. The display of art

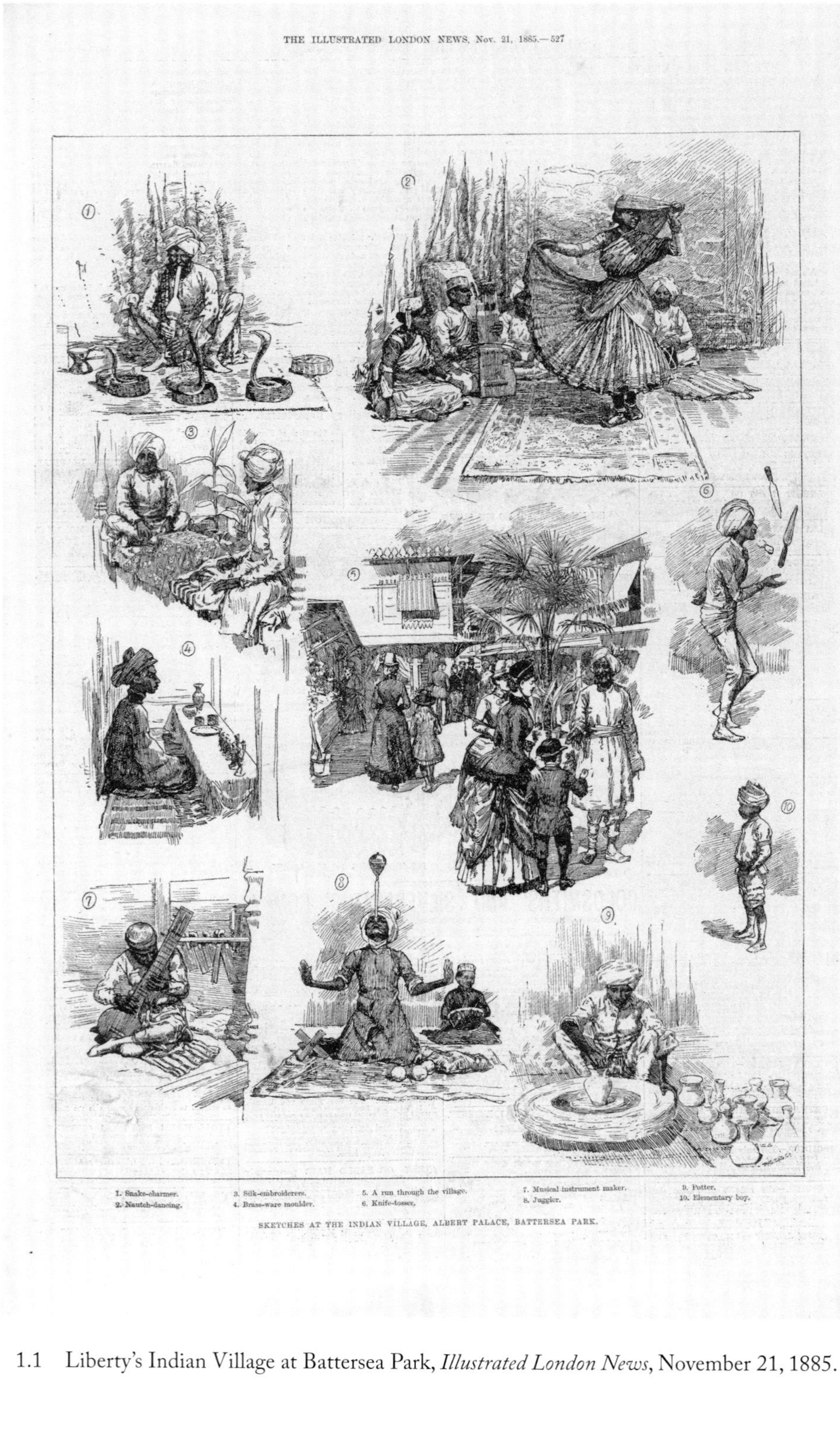
THE ILLUSTRATED LONDON NEWS, Nov. 21, 1885.—527

1. Snake-charmer.
2. Nautch-dancing.
3. Silk-embroiderers.
4. Brass-ware moulder.
5. A run through the village.
6. Knife-tosser.
7. Musical-instrument maker.
8. Juggler.
9. Potter.
10. Elementary boy.

SKETCHES AT THE INDIAN VILLAGE, ALBERT PALACE, BATTERSEA PARK.

1.1 Liberty's Indian Village at Battersea Park, *Illustrated London News*, November 21, 1885.

wares comprised 38 percent of the total Indian Section, while the Palace, Courtyard and Shops occupied a mere 8 percent.[9]

The split between objects displayed according to regional taxonomies and the performing makers was noticeable in location and spatial allocation. The organizers sought to 'bring before us the technical processes of their manufactures, the application and use of objects which otherwise would have remained but beautiful curiosities suitable only for the glass cases of a Museum'.[10] Although there were skilled woodcarvers, a coppersmith, brocade weavers, carpet weavers, a potter and stone carvers active in the Courtyard outside the Palace, visitors continued to be fascinated by the differences between these foreign artisans and contemporary British craftsmen, tending to lump the exhibitors together with the clay figures of tribals in the ethnographic exhibit or the clay figures of soldiers mounted throughout the interior of the Palace's Durbar Hall. An Indian visitor observed that the potter remained motionless and quiet for long periods of time, focused upon his work, and commented that most visitors 'do not see their work, but only look at them and their movements'.[11]

In spite of the presence of skilled makers and adjacent exhibitions of a range of products produced by similar craftsmen, the courtyard of artisans failed to achieve financial success:

> not because the wares manufactured by the artisans were not appreciated, but because the recovery of the cost of the undertaking was throughout considered of less importance than the making of an attractive and picturesque display.[12]

Otherness and the picturesque seemed to be valued more than the skill of the maker or the quality of the product. For example, the queen purchased five paintings of artisans by Rudolf Swoboda, but was more concerned with the faces and ages of the craftsmen than with their products. Only the work of the coppersmith seems to have entered the Royal Collections, and his vessels have received less attention than the gifts of different Indian maharajas and officials.[13] Even the elaborately carved Durbar Room of the 1886 Indian Palace underscores the priorities. The overall scheme, provided by C. Purdon Clarke, the Superintendent of the Indian Museum, featured a strange, dense combination of Mughal, Moorish and Jacobean features, and was carried out by Muhammed Baksch and Mohammed Juma, two Punjab craftsmen who worked upon the interior from June 1885 to April 1886. Thus by the time the Exhibition opened in May, there were no craftsmen active in the Palace interior, the style of which was an imaginary concoction.[14]

While many subsequent exhibitions, such as the Glasgow International Exhibition of 1888, focused upon the products of the 'timeless' traditional art wares of precapitalist India and dispensed with the demonstrating craftsmen, the notion of artisanal display was far from obsolete in London. In 1894, the Hungarian entertainment entrepreneur Imre Kirafly built an exhibition hall in Earls Court in London, creating a large Mughal-style complex with easy links to public transportation. Although he listed George Birdwood and C. Purdon Clarke as advisers, Kirafly seemed less interested in the display of historical objects or regional works of art and more interested in entertainment. For the 1895 Empire of India exhibition and the 1896 India and Ceylon Exhibition, he staged grand India-inspired theatrical works in the Empress Theatre and set up a series of booths where 'native craftsmen pursued their avocations'. These booths, larger than those at the 1886 exhibition, all featured commercial shop-like signs that noted the craftsmen's names, trades and locations.[15]

The Japanese Village

However, the use of the village trope to delineate differences between an imagined exotic traditional culture that made curious objects and a more modern European culture of plentiful machine-produced possessions was not unique to colonial Indian craft. In January 1885, a Japanese Village opened for visitors in Humphreys' Hall, an exhibition space near Albert's Gate in Knightsbridge (Figure 1.2). Within this modern iron and glass exhibition hall was a series of structures, including a Buddhist temple, a tea room and several craft shops. One hundred native rural Japanese, none of them chosen for particular craft expertise, were hired to build the Village and then to populate it. Among the crafts demonstrated in very restricted shop spaces barely more than 3 meters in width were weaving, pottery, woodcarving, lacquering, inlaying of ivory, pearl or metal, lantern-making and fan-making. Visitors commented on how the craftsmen 'squatted' to work rather than standing at a bench, pulled their saws or planes rather than pushed them, and used their feet and bodies to hold their work. The spectators remarked on how the craftsmen worked in:

> the unhasting yet unresting manner which so eminently distinguishes the oriental from the western races. Their long brown hands do not look overly clean, and they hold their tools in what seemed to us an unhandy fashion, but the effects they produce are true and neat and exquisitely finished.[16]

The emphasis remained on craft entertainment, spectacle and difference. 'True and neat and exquisitely finished' was a relative judgment of quality; the products were quaint and surprising given the crude foreign techniques. Among the British visitors who watched 'the dextrous mechanics artistically working' and visually consumed the village were the designer Christopher Dresser and the dramatist W.S. Gilbert.[17]

Nor was the environmental display concept of preindustrial Japanese craft unique to London as an imperial centre. The non-threatening representation of a world of work embedded in what was cast as an unchanged traditional social structure appealed to the racial stereotypes held by a large number of white Western peoples throughout the world; they celebrated their own industrial progress while distancing themselves and appropriating aesthetic principles from 'the other' without contaminating themselves. A traveling Japanese village toured throughout the United States, opening in San Francisco in the fall of 1885 and then traveling to New York City, Boston, Philadelphia, New York again, Chicago, Cincinnati and Baltimore. In New York's Madison Square Garden, the pottery, bronze work and tea preparation caught the eyes of 'fashionable, well-dressed people', and in Boston the village was installed at Horticultural Hall while *The Mikado* was performed at a nearby theatre. As in London, a local importer of Japanese 'curios and wares', Frank Dakin, financed the enterprise. However, when the Village was in

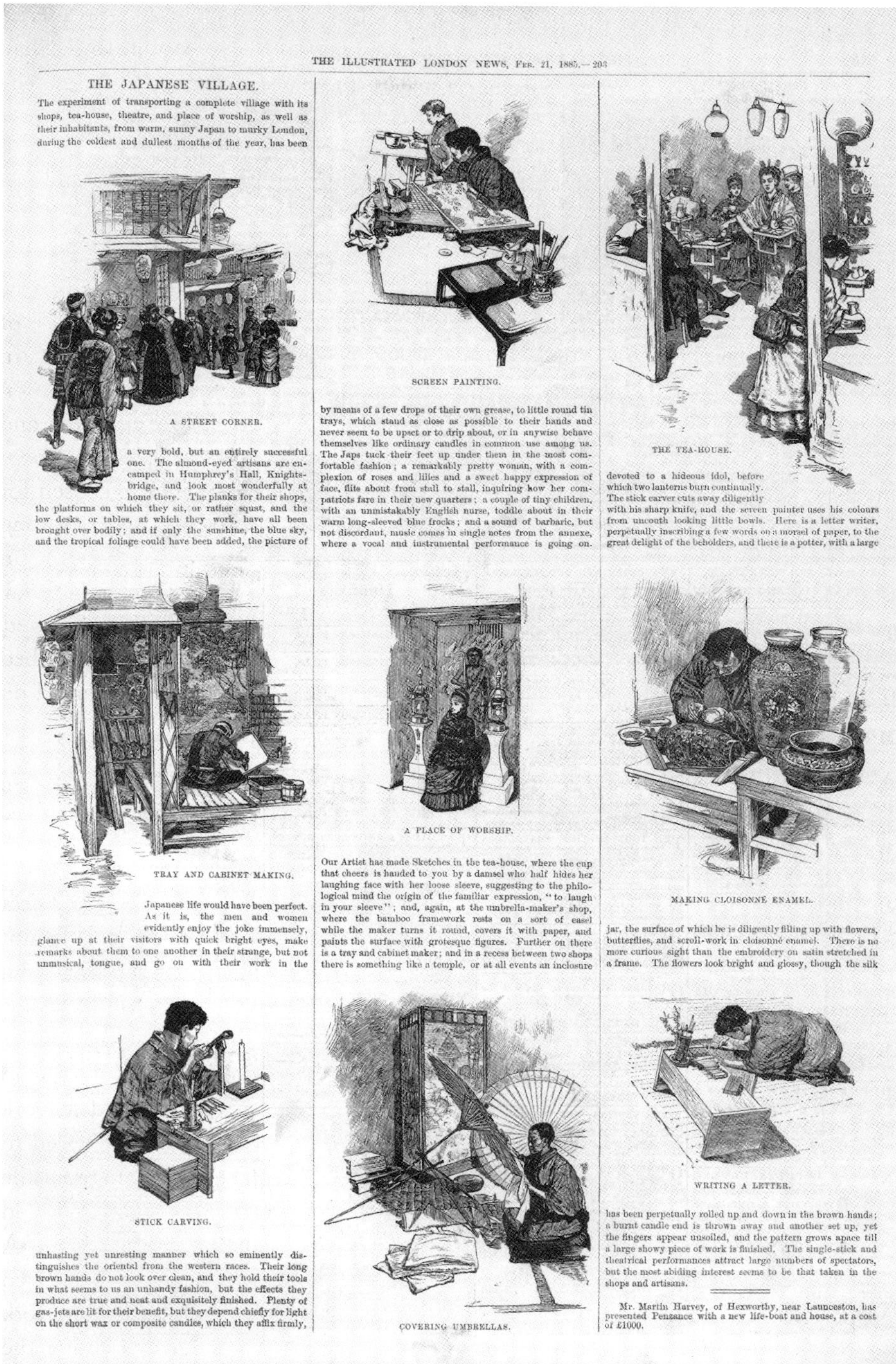

THE ILLUSTRATED LONDON NEWS, Feb. 21, 1885.—203

THE JAPANESE VILLAGE.

The experiment of transporting a complete village with its shops, tea-house, theatre, and place of worship, as well as their inhabitants, from warm, sunny Japan to murky London, during the coldest and dullest months of the year, has been

A STREET CORNER.

a very bold, but an entirely successful one. The almond-eyed artisans are encamped in Humphrey's Hall, Knightsbridge, and look most wonderfully at home there. The planks for their shops, the platforms on which they sit, or rather squat, and the low desks, or tables, at which they work, have all been brought over bodily; and if only the sunshine, the blue sky, and the tropical foliage could have been added, the picture of

TRAY AND CABINET MAKING.

Japanese life would have been perfect. As it is, the men and women evidently enjoy the joke immensely, glance up at their visitors with quick bright eyes, make remarks about them to one another in their strange, but not unmusical, tongue, and go on with their work in the

STICK CARVING.

unhasting yet unresting manner which so eminently distinguishes the oriental from the western races. Their long brown hands do not look over clean, and they hold their tools in what seems to us an unhandy fashion, but the effects they produce are true and neat and exquisitely finished. Plenty of gas-jets are lit for their benefit, but they depend chiefly for light on the short wax or composite candles, which they affix firmly,

SCREEN PAINTING.

by means of a few drops of their own grease, to little round tin trays, which stand as close as possible to their hands and never seem to be upset or to drip about, or in anywise behave themselves like ordinary candles in common use among us. The Japs tuck their feet up under them in the most comfortable fashion; a remarkably pretty woman, with a complexion of roses and lilies and a sweet happy expression of face, flits about from stall to stall, inquiring how her compatriots fare in their new quarters; a couple of tiny children, with an unmistakably English nurse, toddle about in their warm long-sleeved blue frocks; and a sound of barbaric, but not discordant, music comes in single notes from the annexe, where a vocal and instrumental performance is going on.

A PLACE OF WORSHIP.

Our Artist has made Sketches in the tea-house, where the cup that cheers is handed to you by a damsel who half hides her laughing face with her loose sleeve, suggesting to the philological mind the origin of the familiar expression, "to laugh in your sleeve"; and, again, at the umbrella-maker's shop, where the bamboo framework rests on a sort of easel while the maker turns it round, covers it with paper, and paints the surface with grotesque figures. Further on there is a tray and cabinet maker; and in a recess between two shops there is something like a temple, or at all events an inclosure

COVERING UMBRELLAS.

THE TEA-HOUSE.

devoted to a hideous idol, before which two lanterns burn continually. The stick carver cuts away diligently with his sharp knife, and the screen painter uses his colours from uncouth looking little bowls. Here is a letter writer, perpetually inscribing a few words on a morsel of paper, to the great delight of the beholders, and there is a potter, with a large

MAKING CLOISONNÉ ENAMEL.

jar, the surface of which he is diligently filling up with flowers, butterflies, and scroll-work in cloisonné enamel. There is no more curious sight than the embroidery on satin stretched in a frame. The flowers look bright and glossy, though the silk

WRITING A LETTER.

has been perpetually rolled up and down in the brown hands; a burnt candle end is thrown away and another set up, yet the fingers appear unsoiled, and the pattern grows apace till a large showy piece of work is finished. The single-stick and theatrical performances attract large numbers of spectators, but the most abiding interest seems to be that taken in the shops and artisans.

Mr. Martin Harvey, of Hexworthy, near Launceston, has presented Penzance with a new life-boat and house, at a cost of £1000.

1.2 Japanese Village at Knightsbridge, *Illustrated London News*, February 21, 1885.

Cincinnati, Rookwood Pottery hired away Kataro Shirayamadani to work as a ceramic painter.[18]

A privately funded Japanese Village with craft demonstrations was even created during these same years in Australia.[19] Pemberton Willard, a Sydney entrepreneur and theatre producer, promoted a native village that featured Japanese men, women and children 'in native costume' who daily illustrated the 'art industries of Japan'. The accompanying pamphlet praised Japanese craft skill and ingenuity in such products as silks and swords, but also pointed out that a paternalistic government restrained certain modern developments such as rapid locomotion. Furthermore, the Japanese carpenters' drawing of planes towards them or the tailors' stitching away from them seemed odd to the Anglo-Australians, the artisanal equivalent of Japanese writing being oriented from top to bottom and from right to left. The pamphlet explained some of the techniques of cloisonné, fan-making and pottery; but much of the publication was devoted to a discussion of the beautiful motifs found on fans, porcelain, carvings, lacquer work and embroidery. In America and Australia, the emphasis remained on aestheticized entertainment more than on educating a discerning consumer public or raising the artisans' standard of living.

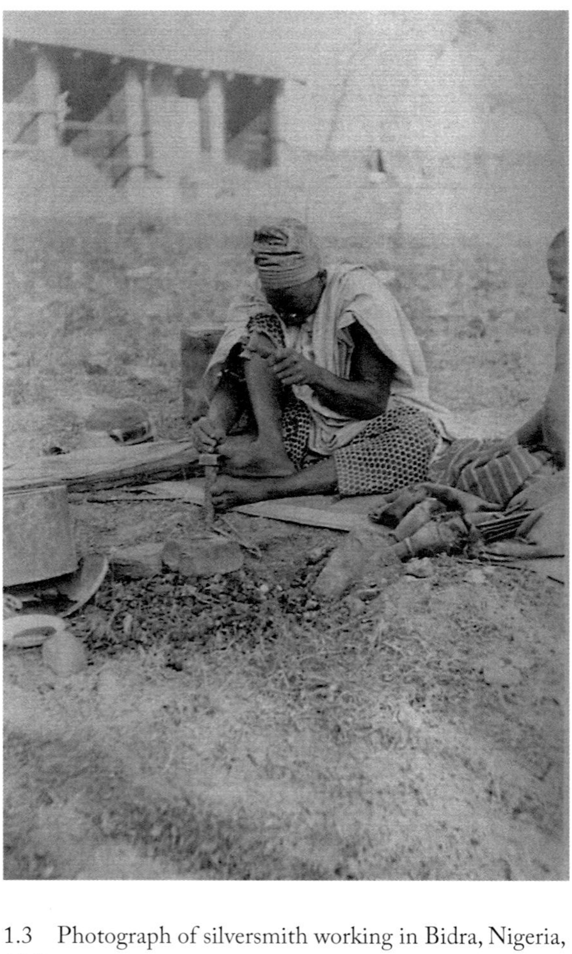

1.3 Photograph of silversmith working in Bidra, Nigeria, 1912.

Twenty years later the curiosity about unfamiliar artisanal landscapes continued to colour Anglo views of craft, but at this point on site rather in the metropole. British visitors to Nigeria and other African parts of the empire were drawn to the sitting work posture, rudimentary tools and low-technology processes characteristic of the African village craftsmen. In terms of an evolutionary schema of stages of production popular since the last quarter of the nineteenth century, African work seemed not only timeless, but also more primitive than the time-clock-monitored and mechanized labour systems of industrial Britain. Thus the British visitors focused once again upon the setting and practice of craft more than the actual products. While a few people purchased blackened earthenware pottery from Ilorin, much of that work could not compete with the British preference at this time for painted art pottery produced in Britain or imported from the art schools in India. Influenced by the trade in colonial postcards, people like Dr John William Scott Macfie took photographs of these anthropological curiosities and assembled photo albums that permitted visual appreciation and memories of a distant village craft from the safety of Britain (Figure 1.3).[20]

Rural Industry

What casts this concept of village craft into sharper relief is an alternative phrase that emerged in the Anglo world at the turn of the previous century – the concept of rural industry. Rooted in the Arts and Crafts movement's critique of the quality and beauty of industrial products and the need to return joy to labour, the concept of the mixed agricultural/artisanal lifestyle gained additional weight in the aftermath of the horrors of World War I's mechanical killing. Citizens and officials began to doubt the unassailability of the modern industrial sector, but did not seek to return to seventeenth- and eighteenth-century craft structures of South or East Asia. Rather they looked to their own past and sought to refashion it as noble and productive.

The term 'rural industry' was first used in the 1890s, but gained particular traction with the establishment of the Rural Industries Bureau in England in 1921 as part of a greater reconstruction effort after World War I.[21] Concerned by the rural stagnation and depopulation in England and Wales, and recognizing the deep value of country life to the English psyche, the government developed a specific agency to maintain and develop viable and attractive rural craftwork. Defining rural industries as activities carried out in home shops or small to mid-sized shops, the Bureau cast a wide net from village trades (blacksmith, wheelwright, saddler, carpenter, turner, basket maker) to seasonal work (making hurdles or fences) to small firms located in particular areas due to cheap power of local materials (textile mills in Wales and Scotland, potteries in the southwest) to outwork (glovemaking, hosiery) to revived work (hand-weaving, leatherwork, rush seats). Publishing small pamphlets, mounting exhibitions, running workshops and providing grants, the Bureau addressed issues of pricing, debt and attitudes towards innovation (Figures 1.4 and 1.5). While the officials recognized that many rural craftsmen had learned skills more in response to the demands of a former age, they emphasized that these skills could be adapted and developed to meet current needs and used government funds to aid this transformation.[22]

The concept of rural industry as a fundamental part of national character and economic policy can also be found in the discussion of the Southern Highlands in the United States. During the Depression of the 1930s, many government officials focused their attention on the people of the Appalachian Mountains, widely accepted as the pure exemplars of American rural life, a living connection to the colonial past and democratic, self-sufficient values. Concerned about the quality and prestige of the 'mountain industries', the government sought to organize the crafts as a distinctive

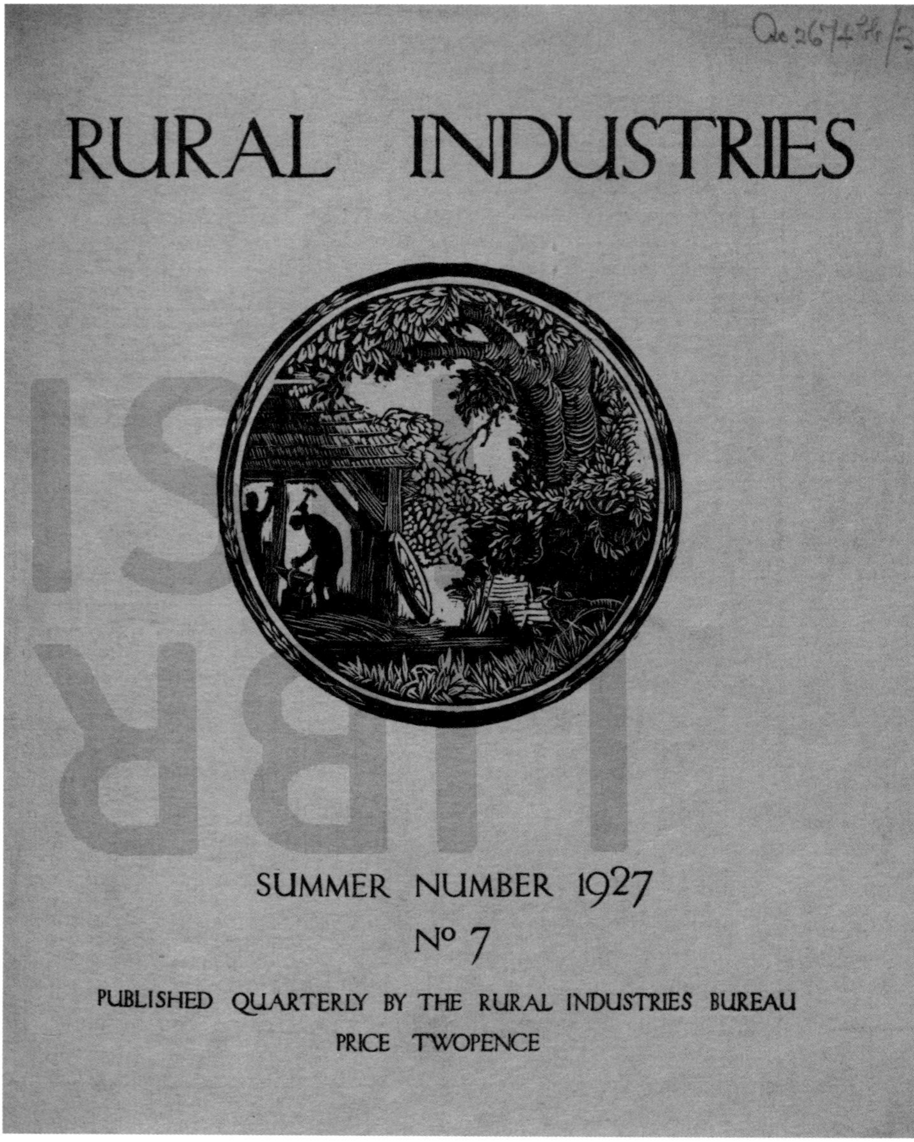

1.4 Cover of *Rural Industries*, 7 (Summer 1927).

industry, developing the cooperative, Southern Highlanders, Inc., to promote adaptation of new business methods and machinery. Like the Rural Industries Bureau, Southern Highlanders, Inc. sought to provide the latest expertise in designing and marketing, and to modernize the distribution of the finished goods. The emphasis was more on the associational aspects of the finished goods and less on the identity of the individual maker or the demonstration of processes. At the same time government agencies such as the Tennessee Valley Authority sought to industrialize mountain crafts and develop retail shops in locations such as the Rockefeller Center in New York City; some

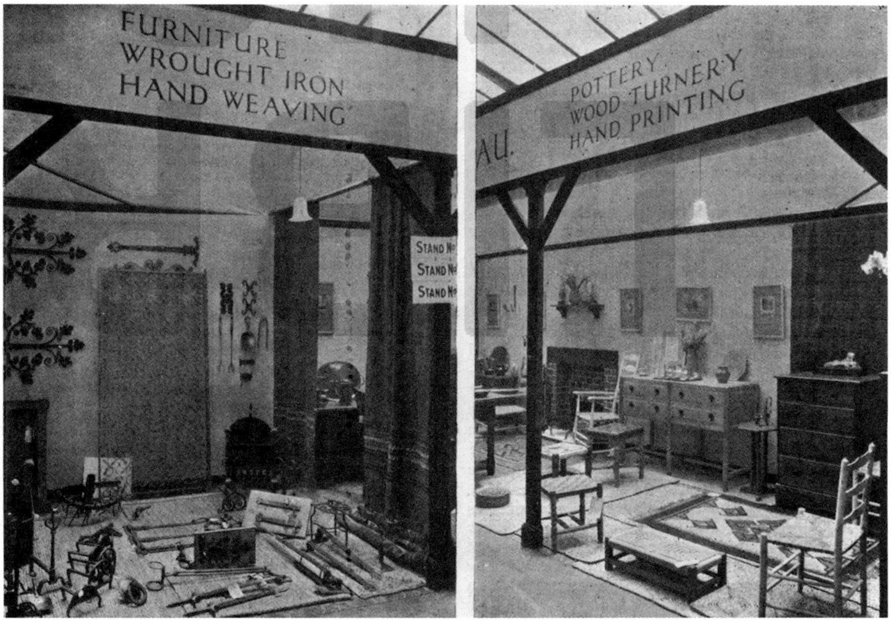

1.5 Photograph of 1927 Rural Industries exhibition.

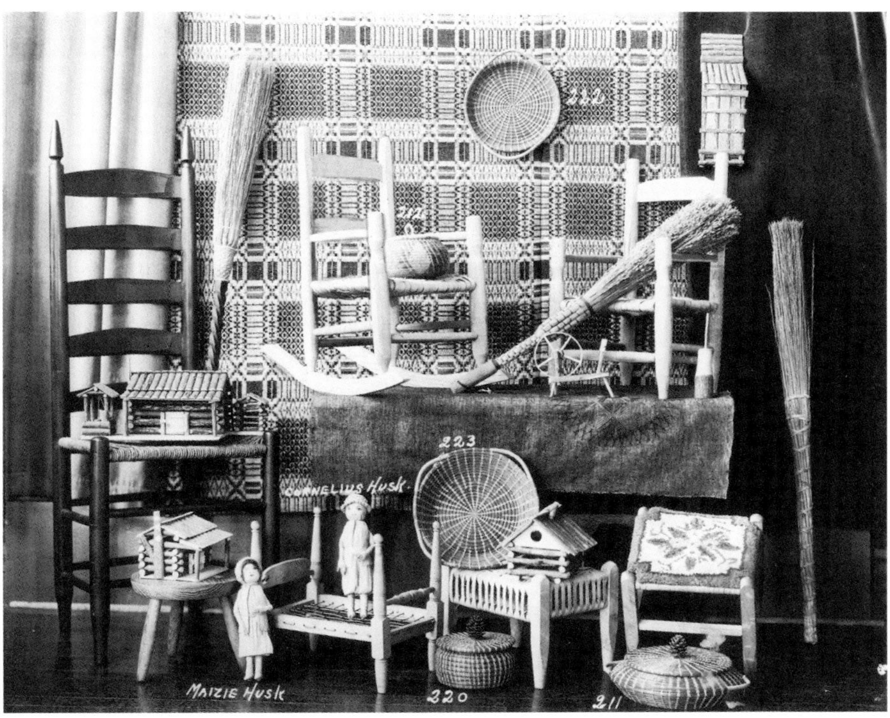

1.6 Photograph of items for sale at the Allanstand Cottage Industries, Asheville, North Carolina, 1930.

local craftsmen instead sought to preserve the 'handicraft culture' by establishing the Southern Highland Craft Guild and intensifying the marketing of handcraft through education and exhibitions. However, the split was more over means than over goals. The same sorts of splint-seated chairs, overshot woven textiles, cornhusk dolls, rustic brooms, pine-needle baskets and kick-wheel-thrown earthenware were sold by both organizations (Figure 1.6). Many people belonged to both organizations, and members of both showed at the 1937 Rural Arts Exhibition in Washington.[23]

Rethinking Village Craft and Rural Industry

In charting the historical discourses of village craft and rural industry, it is easy to theorize the difference along racial lines: the orientalizing gaze of the white colonizer did not allow the acceptance of subaltern colonized artisans as anything but quaint powerless actors in the preservation of an outdated lifestyle. In this narrative, played out in the constructed villages of the nineteenth century or the photographs of the early twentieth century, the artisans were denied any real agency or sense of their role in modernization. They were alienated or abstracted from their actual work and became a form of commodity themselves, more so than their products. In contrast, the rural industries in England and the United States were valorized as essential, vitalizing foundations of a modern state, links between the democratic past and the productive present, with a stake in modern England and America. Their products became morally charged exemplars of this rooted modern perspective and were eagerly consumed. But was the situation that cut and dried, that polarized?

Rather than following this colonial perspective, we might look at the 'experience of incongruity' between the ideological function of industrial capitalism in the place of origin and the material reality in the place of adoption.[24] Instead of focusing upon the erosion of distinctions between core and periphery as capitalism penetrated the marketplace, we should pay attention to the changing combinations of non-capitalist modes of production in relationship to global capitalism. Colonialism is far from a seamless, linear process, but a set of shifting relations that have to be explored from both sides of the colonial structure. As C.L.R. James reminds us in his magisterial book on cricket, a colonizer's culture can be embraced and improved by the colonized to assert their own cultural identity that resists the colonizer's intentions.[25]

The tensions between expectations in London and aspirations in another emerging nation state were easily apparent in the Japanese Village in Kensington. The Japanese government was invested in portraying their culture as advanced, sophisticated and global in the work they sent to the international exhibitions of the late nineteenth century. For example, at the 1878 Paris Universal Exposition, the Kiryu Kosha Kaisha (First Industrial Manufacturing Company) submitted a large elaborate bronze incense burner, consisting of a large lidded vase set upon a tree root with a peacock and peahen also on the base (Figure 1.7). The highly refined object was purchased by Samuel Bing, who then sold it to the South Kensington Museum for the considerable sum of £1,586 *7s. 2d.* in 1883. In contrast, the private English commercial venture behind the

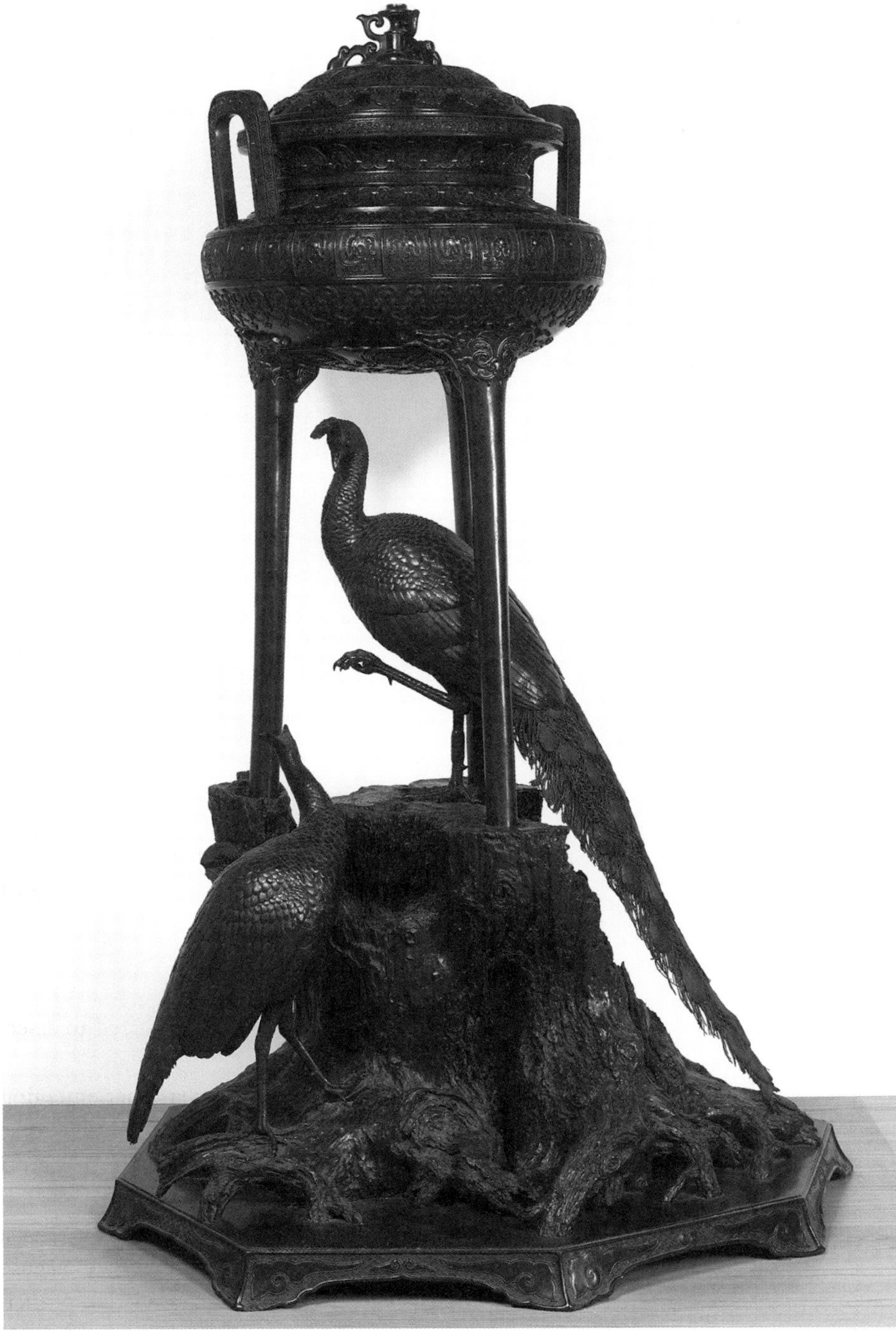

1.7 Koro (incense burner), made by Suzuki Chokichi for Kiryu Kosha Kaisha (First Industrial Manufacturing Company), Japan, 1878; cast and patinated bronze.

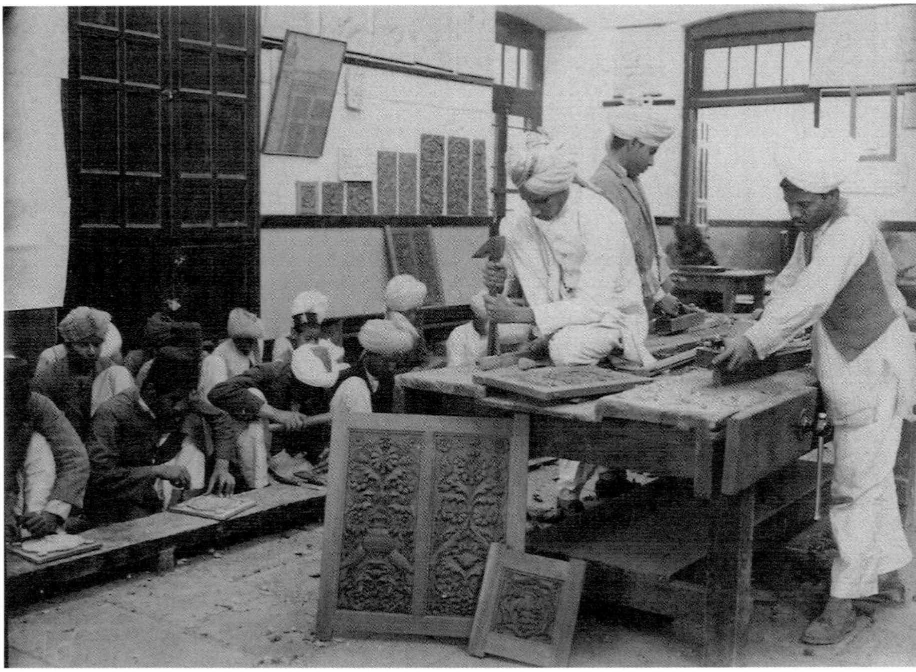

1.8 Photograph of woodworking classroom of the Mayo School of Art, Lahore, 1908.
Note the different levels of working. Some of the students work at a bench, many pupils are working at a low intermediary position with the work on a stool, and no one is working directly on the floor using their body as bench and holdfast. One student follows such traditional practice but he is sitting on the bench.

village portrayed Japan differently, as an isolated, small-scale rural economy. Japanese upper classes and Japanese citizens living in London looked on such a fabricated and marginalized representation unfavourably but could do little to change popular perception.[26]

In regard to colonial India, British cultural leaders who feared that Indian craftsmen might be veering away from their indigenous authentic traditions by incorporating European forms, decoration and materials, developed a notion of 'traditional Indian design' that they promulgated through art schools and exhibitions in India in an effort to keep native traditions alive and untainted by European fashions (Figure 1.8) and to prevent the development of industries that would challenge British manufacturing. In the colonial exchange economy of this era, the Indian craftsmen provided a raw material of artifice – cultural trinkets to inspire British makers and satisfy British consumers seeking authentic souvenirs of the jewel in the crown – but made little work for the local urban market, which instead was to be dominated by British industrial products.

However, the actual craft topography in India was more complex and nuanced. For example, in Jaipur the English surgeon Thomas Hendley drew inspiration from Birdwood's concern about the decline of Indian craft and proposed a Museum of

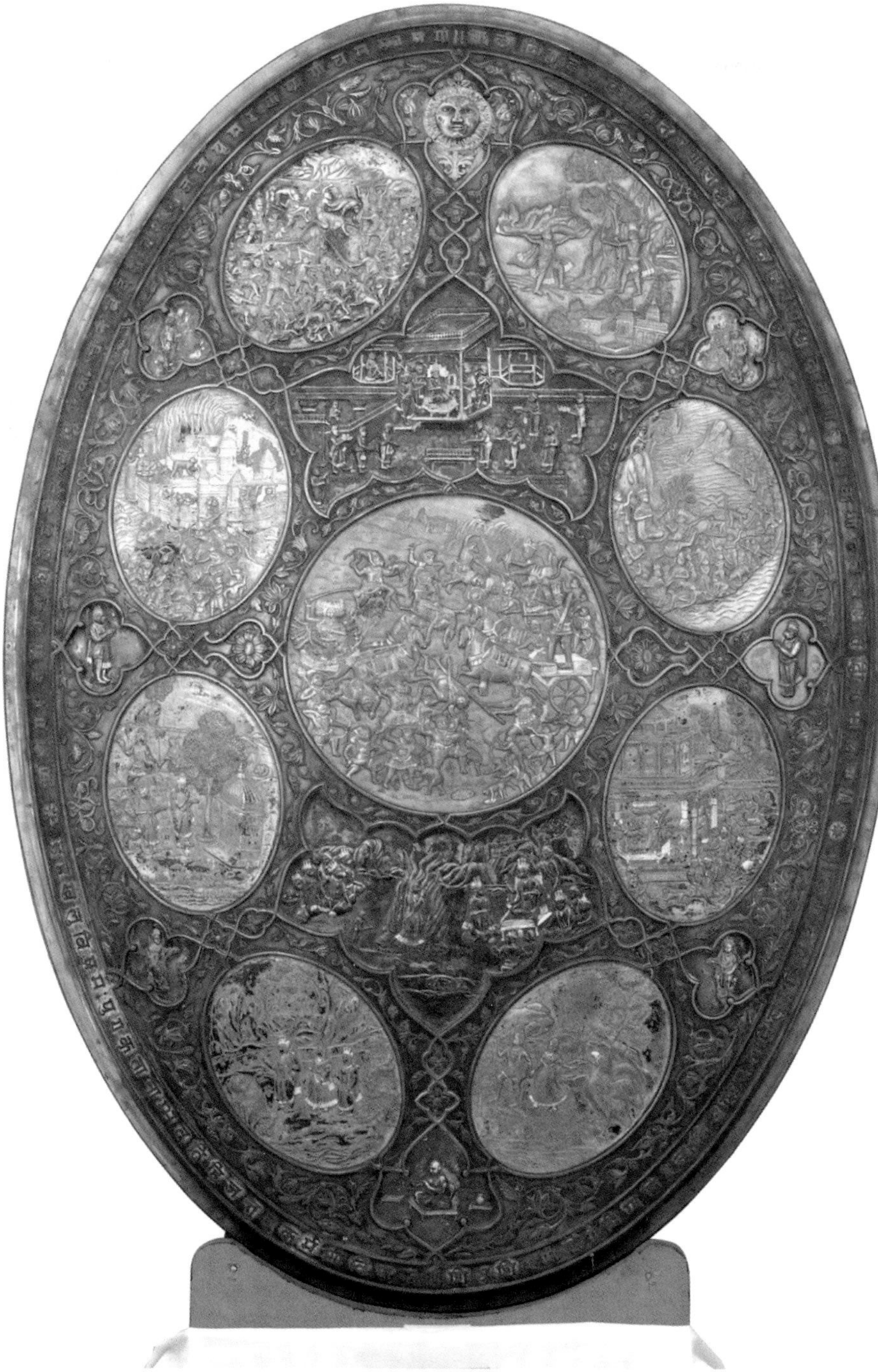

1.9 Ramayana Shield, made by Ganga Baksh, Jaipur, 1885; brass, silver plate.

Industrial Arts (which became Albert Hall) along the lines of the South Kensington model, but it was hardly imposed upon the local community. The Maharajas Ram Singh II and Sawai Madho Singh II established the Jaipur School of Art in 1866 and provided funds to help shape content for the exhibitions and museum of the 1880s. In addition, local administrators such as the school director, the Bengali Opendronath Sen, who developed the concept of blue pottery, and the Jaipur city clerk, Pandit Braj Ballabh, who was the main figure in acquiring the collections of Indian crafts shown at the 1883 Jaipur exhibition and subsequently accessioned in the collection of Albert Hall, played major roles in developing viable regional production and craft sustainability within the new opportunities presented by British officials and tourists and the new emerging Indian middle class.

The new hybrid policy, in which British officials such as Hendley worked with local capital and expertise, was embraced by local artisans, who may have simply considered the British another market, one that provided steady demand and opportunity.[27] These craftsmen actively engaged in it and controlled it, as is suggested by the development of the blue pottery craft in Jaipur. Sen had taken inspiration from the historic fritware tradition of Mughal Delhi and adapted it to the English interest in painted art pottery. At the Jaipur School of Art, potters cast quartz- and feldspar-based fritware clay in plaster molds, developed forms based upon either Indian or European examples, and then painted them in a wide variety of styles that combined Indian landscapes and flowers, Chinese compositions and European borders. Such work contrasted greatly with the more European-influenced work at the Bombay School of Art, which tended to be wheel-thrown earthenware vessels with underglaze decoration. These products catered to Anglo taste for presentation pieces with elaborate modelling and colour or simple Chinese forms with solid coloured glaze.

The work of the Jaipur metalsmith Ganga Baksh, whose Ramayana shield is another distinctively Indian response, in terms of materials, techniques and imagery, was noticeably distinct from the large neoclassical silver and damascened Milton shield made by Elkington & Co. in Birmingham for the 1867 Paris Exposition and from the subsequent electrotype reproductions produced in the 1880s (Figure 1.9). Baksh turned to a sixteenth-century Mughal manuscript as the basis of his images, and relied not simply on repoussé chasing to articulate the imagery, but also on silver and gold plating, damascening and koftkari (the hammering of gilded wire into the brass base).[28]

In the late nineteenth century, the success of some Indian craftsmen in taking advantage of collaboration with English officials and navigating new markets helped foster a new sense of independence that then flowered with the Swadeshi movement. In the first decade of the twentieth century the agency of the Indian craftsman played out even more explicitly when Indian cultural leaders seized the issue of craftsmanship as a tool to overturn British mercantile colonialism and employed the image of the historically prominent craftsman of the seventeenth century, who 'clothed the world', to mobilize Indian nationalism and independence.[29]

Village Craft as Secondary Industry in Interwar West Africa

In West Africa after World War I, British imperial policy sought to modernize its African colonies and help them become more self-sufficient so that they would not drain English capital. The colonial officials resettled native populations from crowded, unhealthy slum environments into new villages praised for their solid construction and spaciousness that allowed good ventilation and firebreaks.[30] The officials also set up English-like schools to develop native bureaucrats and engineers, thereby developing an Anglicized leadership class. In schools like Ugo in Nigeria, Achimoto College in Ghana and Gold Coast Government's School for Senior Boys, craft was used to teach principles of materials and tools, as a form of a science laboratory, and to inculcate Western taste. Coiled, low-fired earthenware cooking and storage pots gave way to wheel-thrown kiln-fired stoneware tablewares, and woodworking increasingly focused upon practical projects like school furniture rather than ritual objects (Figure 1.10). Traditional craft did not have a role in this educational system, but instead new craft exercises were employed to develop a native professional class.[31]

During the economic depression of the 1930s, when the fiscally conservative ideal of a 'trusteeship doctrine' guided the administration of their African colonies, British leaders in the Colonial Office intensified their politicized view of the village ideal. Lumping village craft under the broader category of secondary industries enabled the British government and commercial firms to avoid costly large-scale industrial investment in a region with uncertain prospects for a robust market.[32] But the British leaders were not interested in maintaining traditional and authentic village craft. Instead they sought to preserve the village production units but have the makers produce utilitarian wares for the local and regional markets. Pottery instructors such as Harry Davis at Achimota expanded the students' repertoire to include bricks, tiles and glazed ware, and textile instructors had their students produce extensive yardage.[33]

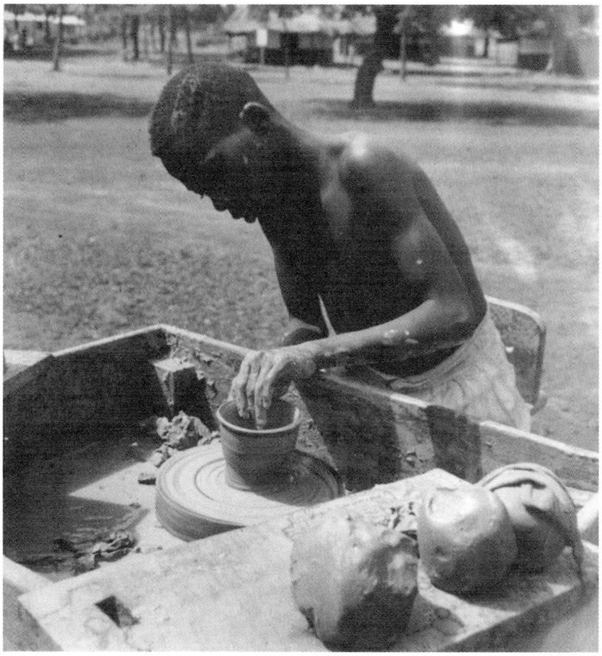

1.10 Photograph of pottery student at Gold Coast Government School, 1930–40.

However, the discussion of village craft was more than pragmatic economic policy; it also served as a means of social control. Village craft would keep the

Africans tied to the past, in familial craft networks based in the village, rather than moving to the cities where they might become too Westernized and develop ideas of individuality and independence, or became unemployed slum dwellers. As R.W. Stopford explained, it was:

> more desirable for social reasons that industrialization at this stage should take the form of developing local craft industries than the creation of a few large manufacturing centres which would accelerate the break-up of village life and weaken the position of agriculture which is the only basis for a satisfying community life.[34]

In response to these attempts to modernize the villages in the 1930s, arts educators like H.V. Meyerowitz and K.C. Murray took more of a preservationist stance and argued for the socially redemptive value of indigenous crafts through the revival of the concept of village craft with skilled craftsmen as the teachers. Critiquing the deplorable quality of European-based arts instruction in African schools in the 1920s and 1930s, which he attributed to pandering to poor European taste and students merely replicating imports, Meyerowitz considered the crafts as 'community art' and praised the local makers for their 'highly developed sense of form, colour, and understanding for materials' that sprang from their own particular communal tribal system.[35] Like George Birdwood in India, Meyerowitz praised the rootedness of native craftsmen and saw the connection between quality craft and mixed agricultural economies. The range and beauty of village products made them superior in function, quality and aesthetics to work produced in the local schools or colleges under European direction.[36]

Murray further detailed the connection of the makers to their village communities: they were naturally admired for their skill, their work served a therapeutic role in an overly impersonal hygienic and utilitarian civilization, and their work with local materials and local clients integrated them into the local economic systems and contributed to social cohesion and cooperative outlooks.[37] Inspired by the work of Meyerowitz, the Rhodes–Livingstone Museum set up an open-air museum called the African Craft Village in northern Rhodesia in 1934. The village's mission was to preserve and promote traditional native crafts, to provide visitors with a full experience of imagined tribal life, and to provide authentic native products for local and distant markets. The promotional and marketing aspect of the venture was clear in a 1958 pamphlet that emphasized the words 'tradition' and 'village' repeatedly, and advertised cameras, lodging and transportation to and from the site.[38]

Both groups of Africanists deployed the concept of village craft in a manner different from the display orientation of the nineteenth century. In the 1930s, they saw craft training and the establishment of native craft centres as the critical issues in the reestablishment of economic welfare and social stability in the African colonies. The village thus moved from entertaining backdrop to economic centrality, from voyeuristic spectacle to pragmatic policy. To Meyerowitz, Murray and others, the idealistic introduction of new industries, social reconstruction and market efficiency had no relationship to the rhythms and values of African life. Sounding like proponents of 'rural industries', they called for policymakers to 'recognize virtues in African culture and to plan progress where most Africans live: that is in the villages and rural towns, rather than to develop further Europeanized towns'.[39]

Conclusion

Just as the nature of village craft is more complicated than a simple racialized colonial perspective argues, so rural industry can be viewed as something other than a heroic effort of regeneration. In Britain, the tendency of the Bureau to focus upon the individual, and narrowly define what rural industry was and what it was not, disconnected it from any possibility of being a complementary but integral form of production. In fact craft was set off in opposition to the production of necessary goods. Instead, a moralizing emphasis and a class hierarchy marginalized rural industry as a form of cultural heritage that simply appeared more modern than village craft, but which was fundamentally quite similar. The Bureau was ultimately in the business of preservation and imagined craft rather than in effective economic policy.[40] The same sort of semantical restriction closed down possibilities in the Southern Appalachians and resulted in a dependence upon collection and fetishization of archetypal craft objects and the canonization of certain craftsmen. The main focus became the tourist and collector market for juried Southern Highlands objects rather than regional economic policy.[41]

The marketing of crafts made by craftsmen of lower economic status working in the village or the countryside raises the spectre of double-edged philanthropy. Arthur Morgan, an American who championed rural industry in the Southern Highlands in the 1930s, warned about the pernicious effect of 'camouflaged charity', in which craftsmen were meant to feel economic self-sufficiency while consumers could feel like they were helping mountain craftsmen. Morgan questioned the tendency to 'subsidize mountain people to make things for well-to-do people to buy for less than what they cost'.[42] The asymmetry of this relationship, in which the maker is kept in a subservient position by the constrictions of the subsidy, whether real or indirect, and purchasers affirm their status through a distinctive handmade object, is the conundrum of craft in a modern world. Paternalism, whether in Battersea Park in 1885, West Africa in 1912, Nigeria in the 1930s, or Ashville, North Carolina in the Depression, plays upon this asymmetry.

The long and complicated history of 'village craft' and 'rural industry' underscores the potentially positive *and* sinister aspects of globalized craft. Scholars of material culture and craft have a particularly important role in exploring these nuances. Not content simply with the representation of the artisan at work, they can provide valuable insight into the choice of materials, techniques and motifs to assess the agency of the maker. Study of actual objects and the relationship of these objects to other related forms and types of the region and time period permit the recovery of intent and performance. Through examination of artifacts, a commitment to fieldwork, and listening to the needs and desires of the makers themselves in their own words, we can better understand the modalities of the 'experience of incongruity' in historical and contemporary narratives of globalization.

Manuscript Sources Cited

British Library, Photo Album of the 1896 India and Ceylon Exhibition (Shelf Mark Photo 888).

University Library, Cambridge University, Institute of Education Collection in the Royal Commonwealth Society Collection.

University Library, Cambridge University, J.W.S. Macfie West African Collection in the Royal Commonwealth Society Collection.

Printed Primary Sources Cited

Birdwood, George, *The Industrial Arts of India* (London: Chapman & Hall, 1880).

Brochure for Ten Thousand Villages (2009).

Colonial and Indian Exhibition 1886: Official Catalogue (London: William Clowes and Sons, 1886).

Green, John Little, *The Rural Industries of England* (London: E. Marlborough & Co., 1895).

Guide to the Rhodes–Livingstone Museum Craft Village (Livingstone, Northern Rhodesia: Rhodes–Livingstone Museum, c. 1958).

Illustrated London News, 1885–86.

'The Institute of West African Arts, Industries, and Social Sciences', *Man*, 43 (September–October 1943), pp. 112–14.

Meyerowitz, H.V., 'African Art in Relation to Village Life', in Village Education in Africa: Report of the Interterritorial 'Jeanes' Conference Held in Salisbury, Southern Rhodesia on 27th May to 6th June, 1935 (Lovedale: Lovedale Press, 1936), pp. 274–98.

Meyerowitz, H.V., *A Report on the Possibilities of the Development of Village Crafts in Basutoland* (Morija Printing Works, 1936).

Meyerowitz, H.V., 'Secondary Industries and Inter-Colonial Collaboration', *Africa: Journal of the International African Institute*, 14/4 (October 1943), pp. 177–83.

Murray, K.C., 'Arts and Crafts of Nigeria: Their Past and Future', *Africa: Journal of the International African Institute*, 14/4 (October 1943), pp. 155–63.

The New York Times, 1885.

Pemberton, Willard, *Japanese Village* (Sydney: P.W. Willard, 1886).

Penny Illustrated Paper, 1885.

Royle, J.R., Report on the Indian Section of the Colonial and Indian Exhibition (London: William Clowes and Son, 1887).

The Rural Industries Bureau: Its Objects & Work (London: Rural Industries Bureau, 1926).

Stopford, R.W., 'Some Problems Involved in the Development of Secondary Industries in West Africa', and Theresa Cahan, 'Secondary Industries for Tropical Africa', *Africa: Journal of the International African Institute*, 14/4 (October 1943), pp. 165–69 and 170–76.

The Times, 1885–96.

Westermann, Diedrich, *The African To-Day* (London: Oxford University Press, 1934).

Secondary Sources Cited

Becker, Jane, *Selling Tradition: Appalachia and the Construction of an American Folk, 1930–1940* (Chapel Hill, NC: University of North Carolina Press, 1998).

Beevers, David, '"From the East Comes Light": Two Relics of the Raj in Sussex', *Apollo*, 152 (November 2000), pp. 11–17.

Brandimarte, Cynthia, 'Japanese Novelty Stores', *Winterthur Portfolio*, 26/1 (January 1991), pp. 1–25.

Butler, L.J., *Industrialization and the British Colonial State: West Africa, 1939–1951* (London: Frank Cass, 1997).

Bryant, Julius, 'Royal Gifts from Victorian India: The New Display in the Durbar Room at Osborne House', *Collections Review*, 4 (2003), pp. 116–24.

Chattopadhyay, Swati, 'A Critical History of Architecture in a Post-Colonial World: A View from Indian History', *Architronic*, 6/1 (May 1997). Available at: http://corbu2.caed.kent.edu/architronic/v6n1/v6n1.05a.html.

Cleary, Joe, 'Irish Studies, Colonial Questions: Locating Ireland in the Colonial World', in *Outrageous Fortune: Capital and Culture in Modern Ireland* (Dublin: Field Day Publications, 2007), pp. 14–46.

Cooke, Edward S. Jr, 'Modern Craft and the American Experience', *American Art*, 21/1 (Spring 2007), pp. 2–9.

Dever, Maryanne, *Australia and Asia: Cultural Transactions* (Honolulu, HI: University of Hawaii Press, 1997).

Doris, David, 'Destiny World: Textile Casualties in Southern Nigeria', *African Arts*, 39/2 (Summer 2006), pp. 36–47 and 86–7.

Douglas, Mary, 'When is a Teapot Not a Teapot?', *American Art*, 21/1 (Spring 2007), pp. 19–23.

Friedman, Thomas, *The World is Flat: A Brief History of the Twenty-first Century* (New York: Farrar, Straus and Giroux, 2005).

Greenhalgh, Paul, *Ephemeral Vistas: The Exposition Universelles, Great Exhibitions, and World's Fairs, 1851–1939* (Manchester: Manchester University Press, 1988).

Harrod, Tanya, 'Commentary' and 'Overseas Education and Arts of West Africa', *The Journal of Modern Craft*, 3/1 (March 2010), pp. 89–105.

Harrod, Tanya, *The Last Sane Man: Michael Cardew* (New Haven, CT: Yale University Press, 2012).

Haynes, Douglas, 'Artisan Cloth-Producers and the Emergence of Powerloom Manufacture in Western India, 1920–1950', *Past and Present*, 172 (August 2001), pp. 170–98.

James, C.L.R., *Beyond a Boundary* (New York: Pantheon, 1983).

Khera, Dipti, 'Engraved Epics: Ornamented Metal Objects', in Chandramani Singh (ed.), *Treasures of the Albert Hall Museum Jaipur* (Usmanpura, Ahmedabad: Mapin Publishing, 2009), 58–67.

Mathur, Saloni, *An Indian Encounter: Portraits for Queen Victoria* (London: National Gallery Company, 2002).

Mathur, Saloni, *India by Design: Colonial History and Cultural Display* (Berkeley, CA: University of California Press, 2007).

McGowan, Abigail, '"All That is Rare, Characteristic or Beautiful": Design and the Defense of Tradition in Colonial India, 1851–1903', *Journal of Material Culture*, 10/3 (November 2005), pp. 263–87.

Murray, Kevin, 'Exogamy in World Craft: A South–South Perspective', *Journal of Modern Craft*, 3/2 (July 2010), pp. 137–42.

Okpoko, Ikechukwa, and Uche Ezeadichie, 'Nigerian Traditional Crafts and Sustainable Development', in L.D. Walu (ed.), *Nigerian Traditional Crafts for Self-Reliance* (Jos, Nigeria: National Commission for Museums and Monuments, 2004), pp. 149–70.

Parthasarathi, Prasannan, *Why Europe Grew Rich and Asia Did Not: Global Divergence, 1600–1850* (Cambridge, UK: Cambridge University Press, 2011).

Riello, Giorgio, and Prasannan Parthasarathi (eds), *The Spinning World: A Global History of Cotton Textiles, 1300–1850* (Oxford: Oxford University Press, 2009).

Riello, Giorgio, and Tirthankar Roy (eds), *How India Clothed the World: The World of South Asian Textiles, 1500–1850* (Leiden and Boston, MA: Brill Publishers, 2009).

Rivoli, Pietra, *The Travels of a T-Shirt in the Global Economy* (Hoboken, NJ: John Wiley & Sons, 2005).

Roy, Tirthankar, *Traditional Industry in the Economy of Colonial India* (Cambridge: Cambridge University Press, 1999).

Scholtz, Amelia, '"Almond-Eyed Artisans"/"Dishonoring the National Polity": The Japanese Village Exhibition in Victorian London', *Japanese Studies*, 27/1 (May 2007), pp. 73–85.

Schwartz, Roberto, *Misplaced Ideas: Essays in Brazilian Culture* (London: Verso, 1992).

Snyder, Rachel, *Fugitive Denim* (New York: W.W. Norton & Co., 2008).

Tebutt, Holly, 'Industry or Anti-Industry? The Rural Industries Bureau: Its Objects and Work' (MA Thesis, Victoria & Albert–Royal College Art Program in Design History, 1990).

Tillotson, Giles, 'The Jaipur Exhibition of 1883', *Journal of the Royal Asiatic Society*, Series 3, 14/2 (2004), pp. 111–26.

Trapp, Ken, 'Rookwood and the Japanese Mania in Cincinnati', *Cincinnati Historical Society Bulletin*, 39 (Spring 1981), pp. 64–8.

Williams, William Morgan, *The Country Craftsman: A Study of Some Rural Crafts and Rural Industries Organizations in England* (London: Routledge & Kegan Paul, 1958).

Wintle, Claire, 'Model Subjects: Representations of the Andaman Islands at the Colonial and Indian Exhibition, 1886', *History Workshop Journal* 67/1 (Spring 2009), pp. 194–207.

Notes

1 On the economy, see Thomas Friedman, *The World is Flat: A Brief History of the Twenty-first Century* (New York: Farrar, Straus and Giroux, 2005); and Pietra Rivoli, *The Travels of a T-Shirt in the Global Economy* (Hoboken, NJ: John Wiley & Sons, 2005). On African cloth, see David Doris, 'Destiny World: Textile Casualties in Southern Nigeria', *African Arts*, 39/2 (Summer 2006), pp. 36–47 and 86–7. A similar book that addresses the production and trade of the globalized economy is Rachel Snyder, *Fugitive Denim* (New York: W.W. Norton & Co., 2008).

2 Ikechukwu Okpoko and Uche Ezeadichie, 'Nigerian Traditional Crafts and Sustainable Development', in L.D. Walu (ed.), *Nigerian Traditional Crafts for Self-Reliance* (Jos, Nigeria: National Commission for Museums and Monuments, 2004), pp. 149–70; and Kevin Murray, 'Exogamy in World Craft: a South–South Perspective', *Journal of Modern Craft*, 3/2 (July 2010), pp. 137–42. See also Murray's 'Craft Unbound' blogs for March 9, August 16, October 21 and November 9, 2009. Available at: http://www.craftunbound.net.

3 2009 brochure for Ten Thousand Villages.

4 On the notion of modern craft, see Edward S. Cooke, Jr, 'Modern Craft and the American Experience', *American Art*, 21/1 (Spring 2007), pp. 2–9.

5 The term 'great divergence' was proposed by Kenneth Pomeranz and has become a major factor in the re-examination of long-term East/West economic relations, see *The Great Divergence: Europe, China, and the Making of the Modern World Economy* (Princeton, NJ: Princeton University Press, 2000). Examples of the recent literature addressing the former dominance of Asian craft traditions and the way in which England and the West developed the concept of unskilled Asian craftsmen and undercut their productive capabilities include Giorgio Riello and Prasannan Parthasarathi (eds), *The Spinning World: A Global History of Cotton Textiles, 1300–1850* (Oxford: Oxford University Press, 2009); Giorgio Riello and Tirthankar Roy (eds), *How India Clothed the World: The World of South Asian Textiles, 1500–1850* (Leiden and Boston, MA: Brill Publishers, 2009); and Prasannan Parthasarathi, *Why Europe Grew Rich and Asia Did Not: Global Divergence, 1600–1850* (Cambridge: Cambridge University Press, 2011). In contrast to the 'timeless' view of South Asian craftsmen, William Morris used history to create a specific lineage within the Anglo-centric history of production epochs. In particular, see his 'Lesser Arts of Life' (1882) and 'Art and Industry in the 14th Century' (1887).

6 Saloni Mathur, *India by Design: Colonial History and Cultural Display* (Berkeley, CA: University of California Press, 2007), esp. pp. 27–79; and *Illustrated London News*, 87 (November 21, 1885), p. 527.

7 George Birdwood, *The Industrial Arts of India* (London: Chapman & Hall, 1880), p. 319.

8 Ibid., p. 312.

9 J.R. Royle, *Report on the Indian Section of the Colonial and Indian Exhibition* (London: William Clowes and Son, 1887). See also Swati Chattopadhyay, 'A Critical History of Architecture in a Post-Colonial World: A View from Indian History', *Architronic*, 6/1 (May 1997). Available at: http://corbu2.caed.kent.edu/architronic/v6n1/v6n1.05a.html.

10 Colonial and Indian Exhibition 1886: Official Catalogue (London: William Clowes and Sons, 1886), p. 87.

11 Mathur, *India by Design*, p. 67. For more on the clay figures, which were modeled by Jadu Náth Pal of Krishnagar, see Claire Wintle, 'Model Subjects: Representations of the Andaman Islands at the Colonial and Indian Exhibition, 1886', *History Workshop Journal*, 67/1 (Spring 2009), pp. 194–207.

12 Royle, Report on the Indian Section, p. 14.

13 The portraits were of Ruxshiram, the 102-year-old potter from Agra; Shaban, the 45-year-old silk spinner and gold and silver weaver from Benares; Radhabulllub Mistry, the 38-year-old carver in marble and stone from Maradburon; Ramal, the 9-year-old carpet weaver from Gambur; and Mohammed Hosein, the 26-year-old coppersmith from Delhi. Hosein's repoussé chased and pierce vessels are now placed on the mantle in the Durbar Room of Osborne House, but they have not always been there. See Saloni Mathur, *An Indian Encounter: Portraits for Queen Victoria* (London: National Gallery Company, 2002) and Julius Bryant, 'Royal Gifts from Victorian India: The New Display in the Durbar Room at Osborne House', *Collections Review*, 4 (2003), pp. 116–24.

14 David Beevers, '"From the East Comes Light": Two Relics of the Raj in Sussex', *Apollo*, 152 (November 2000), pp. 11–17.

15 *The Times* (April 18, 1896); Photo Album of the India and Ceylon Exhibition (British Library, Shelf Mark Photo 888); and Paul Greenhalgh, *Ephemeral Vistas: The Exposition Universelles, Great Exhibitions, and World's Fairs, 1851–1939* (Manchester: Manchester University Press, 1988).

16 See *The Times* (January 10, 1885); *Penny Illustrated Paper* (January 17, 1885); and *Illustrated London News*, 86 (February 21, 1885), p. 203.

17 Amelia Scholtz, '"Almond-Eyed Artisans"/"Dishonoring the National Polity": The Japanese Village Exhibition in Victorian London', *Japanese Studies*, 27/1 (May 2007), pp. 73–85.

18 *The New York Times* (December 4, 1885); Ken Trapp, 'Rookwood and the Japanese Mania in Cincinnati', *Cincinnati Historical Society Bulletin*, 39 (Spring 1981), pp. 64–8; and Cynthia Brandimarte, 'Japanese Novelty Stores', *Winterthur Portfolio*, 26/1 (January 1991), pp. 1–25.

19 A Japanese Village opened in Sydney in April 1886, and in Melbourne in July 1886: Pemberton Willard, *Japanese Village* (Sydney: P.W. Willard, 1886); and Maryanne Dever, *Australia and Asia: Cultural Transactions* (Honolulu, HI: University of Hawaii Press, 1997).

20 See the images and captions preserved in the J.W.S. Macfie West African Collection in the Royal Commonwealth Society Collection at the University Library, Cambridge University. A specialist in tropical medicine, Macfie was a member of the West African Medical Service from 1910 to 1922. He photographed the landscape, architecture, marketplaces and craft activity.

21 An early use of the term is John Little Green, *The Rural Industries of England* (London: E. Marlborough & Co., 1895). See also the books on rural crafts written in the first two decades of the twentieth century by George Bourne, the nom de plume of George Sturt.

22 *The Rural Industries Bureau: Its Objects & Work* (London: Rural Industries Bureau, 1926); and William Morgan Williams, *The Country Craftsman: A Study of Some Rural Crafts and Rural Industries Organizations in England* (London: Routledge & Kegan Paul, 1958). The Rural Industries Bureau continued to publish books on blacksmithing, furniture making, thatching and business practices into the 1950s and even the 1980s.

23 For a thorough discussion of the rural industries in Appalachia, see Jane Becker, *Selling Tradition: Appalachia and the Construction of an American Folk, 1930–1940* (Chapel Hill, NC: University of North Carolina Press, 1998); and *Southern Arts and Crafts, 1890–1940* (Charlotte, NC: Mint Museum of Art, 1996), pp. 78–112. On Southern Highlanders, Inc., see their c. 1935 catalogue for the New York shop, in the collections of the Southern Highland Craft Guild, Asheville, NC.

24 The phrase 'experience of incongruity' is drawn from Joe Cleary, 'Irish Studies, Colonial Questions: Locating Ireland in the Colonial World', in *Outrageous Fortune: Capital and Culture in Modern Ireland* (Dublin: Field Day Publications, 2007), pp. 14–46; and Roberto Schwartz, *Misplaced Ideas: Essays in Brazilian Culture* (London: Verso, 1992).

25 C.L.R. James, *Beyond a Boundary* (New York: Pantheon, 1983).

26 See Scholtz, '"Almond-Eyed Artisans"/"Dishonoring the National Polity"'.

27 On the hybridity of Jaipur, see Giles Tillotson, 'The Jaipur Exhibition of 1883', *Journal of the Royal Asiatic Society*, Series 3, 14/2 (2004), pp. 111–26.

28 Baksh and his work are featured in Dipti Khera, 'Engraved Epics: Ornamented Metal Objects', in Chandramani Singh (ed.), *Treasures of the Albert Hall Museum Jaipur* (Usmanpura, Ahmedabad: Mapin Publishing, 2009), pp. 58–67.

29 On the concept of deindustrialization and the rise of artisanal nationalism, see Mathur, *India by Design*; Abigail McGowan, '"All That is Rare, Characteristic or Beautiful": Design and the Defense of Tradition in Colonial India, 1851–1903', *Journal of Material Culture*, 10/3 (November 2005), pp. 263–87; Tirthankar Roy, *Traditional Industry in the Economy of Colonial India* (Cambridge: University of Cambridge, 1999); and Douglas Haynes, 'Artisan Cloth-Producers and the Emergence of Powerloom Manufacture in Western India, 1920–1950', *Past and Present*, 172 (August 2001), pp. 170–98.

30 The 1939 example of the slums of Anchau and the new village of Takalafiya is documented in the Institute of Education Collection, items 38–42 in the Royal Commonwealth Society Collection at the University Library, Cambridge University.

31 See the images and captions preserved in the Institute of Education Collection in the Royal Commonwealth Society Collection at the University Library, Cambridge University.

32 On the specifics of trusteeship, see L.J. Butler, *Industrialization and the British Colonial State: West Africa, 1939–1951* (London: Frank Cass, 1997). The notion of secondary industries is detailed in R.W. Stopford, 'Some Problems Involved in the Development of Secondary Industries in West Africa', and Theresa Cahan, 'Secondary Industries for Tropical Africa', *Africa: Journal of the International African Institute*, 14/4 (October 1943), pp. 165–9 and 170–76.

33 'The Institute of West African Arts, Industries, and Social Sciences', *Man*, 43 (September–October 1943), pp. 112–14; and Tanya Harrod, *The Last Sane Man: Michael Cardew* (New Haven, CT: Yale University Press, 2012).

34 Stopford, 'Some Problems Involved in the Development of Secondary Industries in West Africa', p. 168; and Diedrich Westermann, *The African To-Day* (London: Oxford University Press, 1934), pp. 93–107.

35 H.V. Meyerowitz, 'African Art in Relation to Village Life', in *Village Education in Africa: Report of the Interterritorial 'Jeanes' Conference Held in Salisbury, Southern Rhodesia on 27th May to 6th June, 1935* (Lovedale: Lovedale Press, 1936), and Meyerowitz, *A Report on the Possibilities of the Development of Village Crafts in Basutoland* (Morija Printing Works, 1936). See also the excerpts of articles from *Overseas Education* and the introductory commentary by Tanya Harrod in *The Journal of Modern Craft*, 3/1 (March 2010), pp. 89–105. I am grateful to Tanya for sharing with me her references to and insights regarding West African village craft.

36 See Meyerowitz, 'Secondary Industries and Inter-Colonial Collaboration', in *Africa: Journal of the International African Institute*, 14/4 (October 1943), pp. 177–83.

37 K.C. Murray, 'Arts and Crafts of Nigeria: Their Past and Future', *Africa: Journal of the International African Institute*, 14/4 (October 1943), pp. 155–63.

38 *Guide to the Rhodes–Livingstone Museum Craft Village* (Livingstone, Northern Rhodesia: Rhodes–Livingstone Museum, c. 1958).

39 Murray, 'Arts and Crafts of Nigeria: Their Past and Future', p. 155.

40 Holly Tebutt, 'Industry or Anti-Industry? The Rural Industries Bureau: Its Objects and Work' (MA Thesis, Victoria & Albert–Royal College Art Program in Design History, 1990).

41 Becker, *Selling Tradition*; and Mary Douglas, 'When is a Teapot Not a Teapot?', *American Art*, 21/1 (Spring 2007), pp. 19–23.

42 On Morgan, see Becker, *Selling Tradition*, p. 111.

2

Souvenir Art, Collectable Craft, Cultural Heritage: The Wendat (Huron) of Wendake, Quebec

Anne de Stecher

Introduction

The nineteenth-century souvenir arts production of the Wendat (Huron) First Nation of Wendake, Quebec contains within it the story of a community. Through their virtuosity and knowledge in a specialized craft, together with their traditions of diplomacy and trade, the Wendat negotiated the economic challenges and pressures of assimilation facing many Aboriginal groups in the nineteenth century.[1] A diverse array of souvenir objects embroidered with floral and figurative motifs in dyed moosehair, from birch-bark cigar cases and glove boxes to deer-hide moccasins and pouches, was sought after by European travellers and tourists. With this internationally known souvenir craft production as an economic mainstay, the Wendat made a successful transition from the fur-trade era, through difficult decades of change, to relative prosperity in the mid-nineteenth century.

Today, numerous examples of Wendat souvenir work can be seen in museums across Europe, Canada and the United States. They are usually viewed by academics, curators and the public in this institutional space, separated from the historical role they played within the Wendat community as part of an 'aesthetic–cultural' complex involving 'different expressive, performative, and ritual elements'.[2] This chapter situates the souvenir craft within such a complex, exploring its role in sustaining community and identity, and at the same time demonstrating the continuity of Wendat traditions as they responded with forethought and strategic planning to the increasing pressures of colonizing settler presence.

Souvenir Art, Collectable Craft

In my discussion of the Wendat souvenir production I use both the words 'craft' and 'art', often defined as separate categories in western visual arts traditions. However, because they are based in European concepts, each of these terms conveys only partial meanings when used in discussion of objects made within Indigenous communities. I suggest that the terms 'craft' and 'art' are both needed to describe works of virtuosity created through specialized technical skills that explore the expressive, aesthetic possibilities of traditional Indigenous media in forms that serve a function and also have ceremonial, spiritual significance.[3] Craft in its meaning of knowledge and skill that is difficult to attain, held by those who are masters of techniques, media and design, is particularly apt in describing the Wendat souvenir works. The exact techniques, stitches and methods of dyeing and tanning made up a specialized field of knowledge. The European understanding of fine art is also conveyed in these works – in the traditional motifs, use of colours, formal design and iconographic meanings based in Indigenous aesthetic and spiritual beliefs. Central to this discussion of Indigenous craft production is an Indigenous worldview, according to which, as Métis curator Marie Bouchard states, art is seen 'not as separate from everyday life, but rather as an integral and necessary component of our lives'.[4] From this perspective, the terms 'craft' and 'art' come together to describe the intricate processes of production, the layers of meaning and the rich creativity of the souvenir works.

The Wendat Souvenir Craft Tradition: The Quebec Contact Zone

The Wendat souvenir craft was a creative outcome of the Quebec contact zone, a product of the rich intercultural exchange between women of Indigenous communities and the French convents of Quebec. Drawing from both these cultural traditions, Aboriginal materials, techniques and forms came together with European pictorial representation in new modes of expression. Earlier studies have focused on European influences in the Wendat souvenir arts. However, in this essay, I build on new directions in the study of Great Lakes Indigenous material culture that foreground the autonomy of Aboriginal cultures. My approach is to situate the souvenir works within the context of the Wendat community's artistic, commercial and diplomatic traditions, through study of the objects themselves. By foregrounding the authority of the Wendat nation in its engagements with European settler society and seeking a perspective of events grounded in the Indigenous voice, this study adds new dimensions to understandings of the Quebec contact zone.

To bring the story of this community and its craft tradition to life I will discuss two remarkable women, Marguerite Vincent La8inonke (1783–1865) and Caroline Gros-Louis (1876–1941). Marguerite Vincent La8inonke,[5] whose Wendat name means 'la jeune fille' (the young girl),[6] also known as 'la femme habile aux travaux d'aiguille' (the woman skilled in needlework), was a force in the Wendake (Lorette) community in the mid-nineteenth century as an artist, in her role in her family's development of the souvenir industry and as a member of a family of hereditary chiefs noted for their diplomatic and political activism.[7] A pair of trays received from the Wendat community by Lord and Lady Elgin around 1852 have been attributed to her (Figure 2.1).[8]

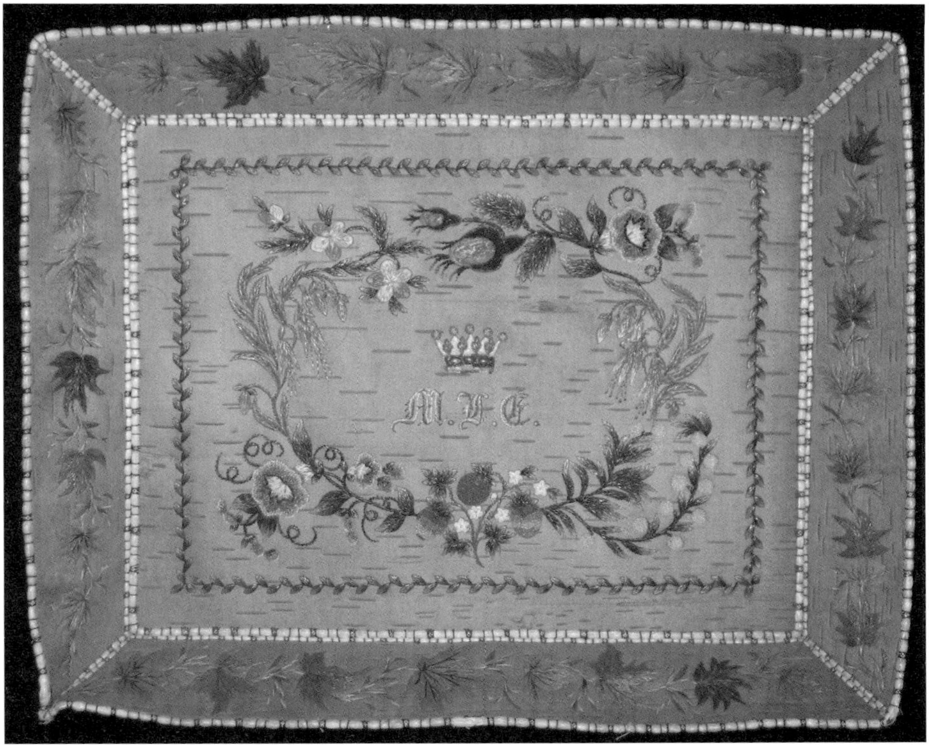

2.1 Moosehair-embroidered birch-bark tray, Elgin Collection, 1847–52.

Caroline Gros-Louis was well known in the early twentieth century for her artistry in moosehair embroidery, particularly for her work in the moosehair-embroidered chief's outfit that was a diplomatic gift to Swedish Consul Folke Cronholm in 1905 (Figure 2.2). Both women were recognized for their technical skills and the artistry of their designs in creating works as commodities, as well as those made for diplomatic and ceremonial use by the community. Their inclusion is an opportunity to expand another important field of study: the identification of individual Indigenous women artists and their oeuvre. The association of known women artists with particular bodies of work is significant. One of the challenges in the study of historical Indigenous arts is that when these works were collected, artists' names were generally not recorded. As a result, the works can usually be studied only by chronology and general style rather than as the creation of specific women. However, in the Wendat craftwork, stylistic analysis brought together with archival research can help to identify different artists.

An important element in the identification of works by individual creators, in the period of souvenir production from the 1830s onward, is recognition of the community of women artists producing works of virtuosity in technique and design, and understanding how traditional knowledge was transmitted. Contemporary Wendat artist Mireille Siouï explained the organization of the souvenir production process in her discussion of the Vincent–Picard family. In the early nineteenth century, they

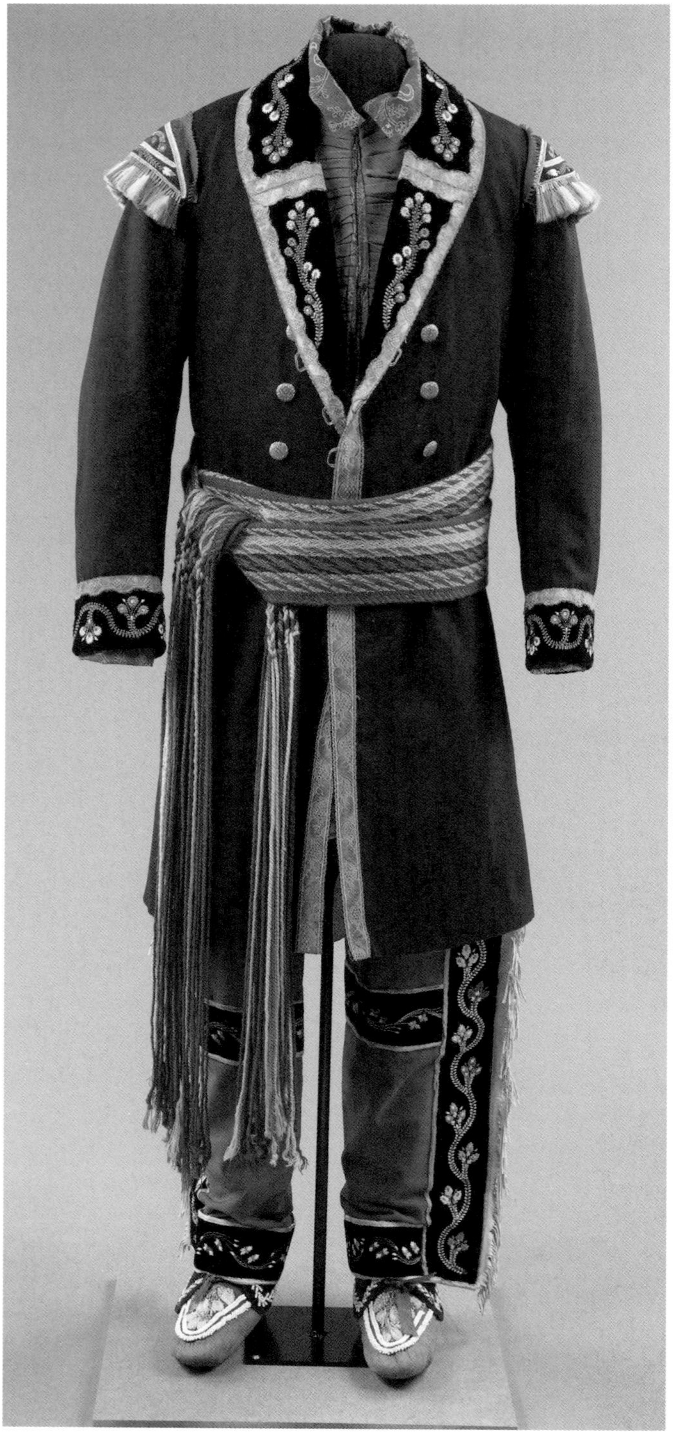

2.2 Caroline Gros-Louis, Chief's outfit, Folke Cronholm Collection, wool, hide, moosehair, feathers, cotton. 1905.

arranged production among a number of families in the community and then supplied souvenir arts to merchants in Quebec, Ontario and New York.[9] Siouï contends that by carrying out these works – sewing, moosehair embroidery and beadwork – at home, Huron women contributed to the transmission of traditional knowledge.[10] While it may be possible through archival research to identify individual artists by name, and through stylistic analysis to identify others, it is important to acknowledge the community of women artists who, over generations, produced and sustained this craft tradition.

The Craft of Moosehair Embroidery

The stories of Marguerite Vincent La8inonke and Caroline Gros-Louis demonstrate the key role played by souvenir craft within the spheres of diplomacy and commerce, and I will situate their narratives within a history of these entwined traditions. I will open this discussion with a brief introduction to moosehair-embroidery techniques

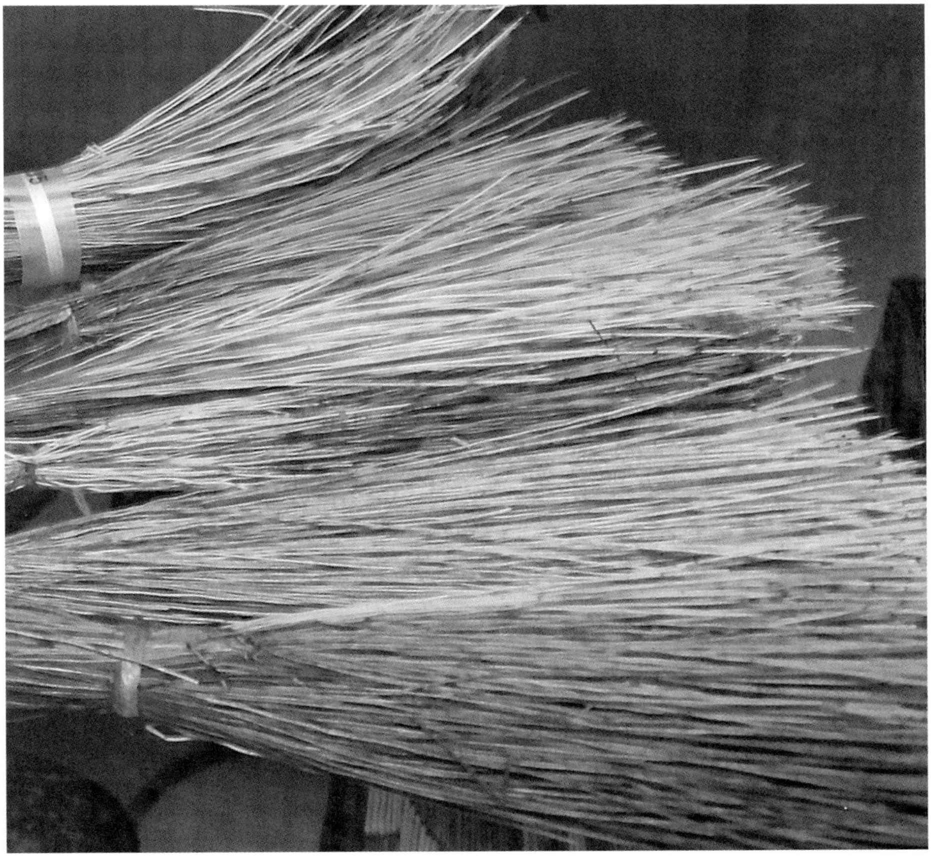

2.3 Dyed moosehair, Hôpital-Général, Quebec City, Quebec.

and design, and thus enable a greater appreciation of the virtuosity of the women who worked in this craft and art.

Moosehair is an embroidery medium with which it is difficult to work. Getting the moosehair itself takes specialized skills; the task of moose hunting was one of Wendat men's important contributions to souvenir production. The moosehair used for embroidery is found at the nape of the neck and the top of the tail. Each strand is about four inches or ten centimetres long. The hair is washed and dyed, originally using vegetable dyes and, later, chemical dyes. The vegetable dyes required specialized knowledge: first, to recognize and locate the plants that provided the colours, in roots, leaves and flowers, and then to prepare them (Figure 2.3). After 1857 and the advent of aniline and chemical dyes, a new range of colours was gradually added to the repertoire, but the use of plant-based dyes continued.

The Wendat souvenirs were made of hide, bark, and bark covered with red or black wool fabric. The Wendat hide moccasins and pouches were known for their dark brown or black colour and velvety texture, achieved through specialized tanning and smoking techniques, requiring recipes for particular bark and wood combinations. For the moosehair embroidery itself, there are two main stitching techniques, illustrated in a pair of moccasins from the collection of the Bata Shoe Museum and the birch-bark Elgin tray from the collection at the Canadian Museum of Civilization (Figure 2.4; see also Figure 2.1). The moccasins demonstrate the appliqué technique usually seen on hide. Several strands of moosehair are laid on the hide and are attached to the surface with sinew or cotton thread to form designs. This appliqué technique is similar to the techniques of working with porcupine quill and was probably in use before European contact.[11]

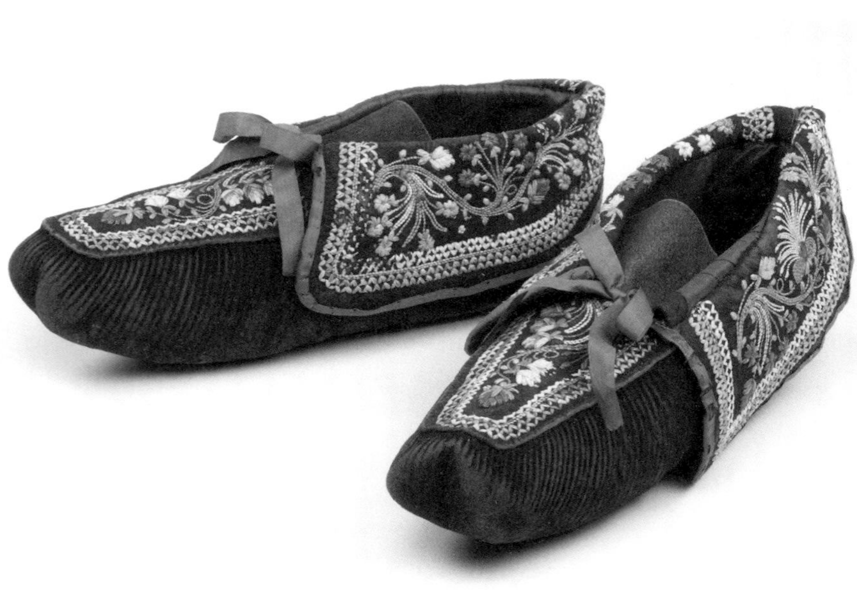

2.4 Moosehair-embroidered hide moccasins, late 1840s.

The Elgin tray demonstrates the embroidery technique used with birch bark; the moosehair is stitched through the bark. In some objects outlines drawn on the bark are visible; they act as guides for the embroidery stitches to follow. From the individuality of each motif, it does not appear that patterns were used. This technique probably developed in the seventeenth century in Quebec convents, through exchange of needlework techniques and materials between the French nuns and the Indigenous girls who attended the convent school.[12] Both techniques require great skill and patience. Lady Simcoe wrote in 1791 that the moosehair was so short, 'it must be put through the needle for every stitch, which makes it tedious'.[13] The body of knowledge, technical ability with difficult materials and design aesthetic necessary to produce exquisite moosehair-embroidered objects made this a craft and art that required patience and dedication. It embodied Wendat cultural traditions and brought meaning, significance and value to the diplomatic ceremonies of which it was a part, and also to European buyers who held the work in such high esteem.

Entwined Traditions: Commerce, Diplomacy and Visual Arts

The Wendat community's history of trade and military alliance with European communities dates from the early seventeenth century. Souvenir craft production was a development of the visual arts traditions that were integral to the Wendat diplomatic practices, and they helped to define and sustain these relations. In his landmark text, *The Children of Aataentsic*, a history of the Wendat to 1660, Bruce Trigger describes their 'embedding of marketing within a framework of political and social relations' and the obligations of generosity between friends and allies, shown through gifts.[14] The Wendat continued this tradition of commerce intertwined with political relations through the eighteenth, nineteenth and early twentieth centuries, often strengthened through real or fictional kinship ties, to gain intermediaries in trade and military alliances.[15] The souvenir craftwork emerged within this framework.

In one of the earliest accounts of the importance of craft within the Wendat framework of political and social relations, French explorer Samuel de Champlain described his negotiation of promises of trade and military support in 1609 with Ochateguins, a leader of the Wendat confederacy.[16] In that year, Champlain fought beside the Wendat in a battle against the Iroquois; he showed bravery and his assistance was instrumental to Wendat victory. Shortly after the battle, the Wendat invited him to come to their country, live with them, and help them always as brothers, which he promised to do.[17] On his return to France following these events, Champlain gave a beautifully woven quilled belt to King Henri IV.[18] Champlain's descriptions of Wendat and Algonquin women's skills in such work suggest that this was a gift given to him by leaders of these closely connected communities.[19] The Wendat traditions of diplomacy and the presentation of valuable gifts to mark events, such as the quilled belt, were a means to form ties with military allies and trade partners. Champlain had proved himself a worthy ally in war; the French were highly desirable trade partners. This, together with the Wendat invitation to Champlain to live with them as their brother and the valuable quilled belt, suggest that the Wendat extended to him the honour of ceremonial adoption.[20] The tradition of adoption to strengthen alliances was shared by other Indigenous Great Lakes nations and the narrative of trader–soldier John Long in 1777 gives a detailed description. The ceremony involved a feast with singing, the gathering of community leaders to smoke the calumet pipe with the person to be adopted, the presentation of valued gifts such as belts of wampum, and the bestowal of a new name.[21] In the nineteenth and twentieth centuries, Wendat diplomatic initiatives built from this tradition continued to solidify military and trade commitments.

Relations between the Wendat and Champlain developed further in both trade and military support. The Wendat were strongly placed within the Indigenous nations' commercial and diplomatic landscape in the Eastern Woodlands region; through their trade and military agreements with the Wendat, Algonquin and Montagnais nations, the French were integrated into this framework.[22] In 1615 Champlain spent a winter in a Wendat village and his account of women's visual arts demonstrates the immediate appeal these works held for Europeans. He described the intricate designs and brilliant reds of the porcupine quillwork on their clothing, the same techniques, materials and designs from which the souvenir craft developed.[23]

The Wendat confederacy dispersed in 1649, not long after Champlain's death in 1635, following prolonged conflict with the Haudenosaunee (Iroquois) and periods of famine. A group of Christian Huron from the confederacy relocated to an area outside Quebec City and formed the community of Lorette, now known by its Wendat name, Wendake. Their descendants developed the souvenir craftworks that became a mainstay of trade production in the nineteenth century. In their new location in close proximity to French Canadian settlements, Wendat diplomatic practices continued. At the heart of these ceremonies was the formation of fictive kinship relations through ceremonial adoption, in models adapted to the shifting political and social landscape of Quebec.

The Eighteenth Century

In the eighteenth century, the Wendat of Wendake continued as French military allies and fur-trade partners. In addition, they developed a new business supplying military equipment to the Quebec garrison, an indication of the Wendat community's ability to adapt to changing circumstances and identify new trade opportunities. The French soldiers who came to the Great Lakes region were not well equipped for life in the northeast, and the Wendake community supplied snowshoes, canoes, paddles and other goods.[24] Indigenous communities expanded the production and sale of finished products to European communities, as well as the raw material of furs. Archival evidence suggests that the fur trade and the production of military equipment were the Wendat community's main commercial focus; however, souvenir works were produced for sale or as gifts to European soldiers and visitors. Many examples survive in museum collections.[25]

Men's and women's complementary work roles were integral to Wendat commercial success.[26] Men's hunting skills and abilities in trade and diplomacy provided furs to sell and the relationships between Indigenous and French traders guaranteed markets. Women's knowledge of Indigenous tanning techniques gave hides their superior quality.[27] Although snowshoe moccasins were not listed in the Provisions de Guerre of the French military, it seems probable that the Wendat supplied these as well as the snowshoes. In the nineteenth century it appears that there was also a shift in work roles: men looked after the tanning and smoking of hides, possibly because women's time was devoted to the moosehair embroidery of souvenirs.[28]

The Wendat traditions of marketing within a framework of political and social relations continued. French texts of the period describe the ceremonies held at Wendake for European visitors – syncretic events that integrated Indigenous diplomatic practices with Christian religious traditions. In the late seventeenth century, Jesuit priests had imported two statues of the Virgin Mary for the church at Wendake, reproductions of statues at shrines in Belgium and Italy.[29] As a result, in the eighteenth century the Wendake village church became a noted pilgrimage destination. This provided an opportunity for the community to continue traditional Wendat diplomatic practices. Visitors of note were welcomed into the community with a military reception held by the warriors, a feast, traditional Wendat singing and an oration by Wendat leaders, which was the focus of the event.[30] The feast furthered connections between Wendat and settler communities, while the oration conveyed civilities and at the same time created a space to raise political issues.[31] Although Wendake was known as a Christian community, the highly visible ceremonies of dance and oration proclaimed through performance the strength of Wendat Indigeneity.

In keeping with Wendat intertwining of diplomacy, trade and military assistance, Wendat warriors actively assisted the French in various conflicts. In the early eighteenth century, the Wendat joined the French in raids on settlements in New England, supporting their allies but at the same time following their own purposes – specifically capturing prisoners to adopt into the community.[32] During the Seven Years War (1756–63) the Wendat were again allies of the French forces and fought beside them on the Plains of Abraham near Quebec City.[33] When the British took over Quebec, Wendat leaders met with the British to sign an agreement to make peace and to assure British protection and free passage for the Wendat. In addition, they were assured the right to practice their religion, their traditions and the freedom to do business with the English garrison.[34] This illustrates the political strengths of Wendat leadership; in the midst of this dramatic shift in power, the chiefs worked strategically to bring about a successful diplomatic realignment and a continued market for their goods.

The story of Marguerite Vincent La8inonke and her family exemplifies the strategic initiatives carried out by Wendat leaders and members of the community to sustain Wendat economic and cultural sovereignty and the part played by the souvenir craft tradition. Marguerite Vincent's father, Louis Vincent Sawantanan, was from a family of Wendat chiefs.[35] From the early 1770s, he and students from other Indigenous nations on the St Lawrence attended Dartmouth College, a Protestant school in New Hampshire.[36] In a very different situation from the residential schools of the nineteenth and twentieth centuries, Indigenous students together with students from New England settler communities received a liberal arts education.[37] Vincent graduated in 1781 as a minister, teacher and interpreter, speaking Wendat, French, English and Mohawk. He was the first member of an Aboriginal community in the Canadas to receive a BA.[38] Sending students to this school was a strategic decision by the chiefs of the St Lawrence nations. When he returned to Wendake, Louis Vincent began work on Wendat land claims and he established a school in the community.[39] His knowledge of English culture and language assisted the Wendat in negotiating the British political and legal systems. This was help the community would need in the difficult decades to come.

The Nineteenth Century

The early nineteenth century saw significant changes in the economic and political situation of the Wendat. In the decades after the War of 1812, the British government no longer saw Indigenous communities as military allies but rather as expenses to the Crown, as occupiers of land that could be used by settlers and businesses, and as peoples to be 'civilized' through assimilation. There was competition for land use as increasing numbers of settlers and lumber companies deforested what had been Wendat hunting territory and game resources diminished. Further, private hunting clubs closed off large sections of land to Aboriginal hunters.[40] As result of these changes, the Wendat lost much of their hunting territory and related sources of income. In addition, the annual gifts of guns, ammunition and blankets to Indigenous nations, which had been the symbols of annual renewal of military allegiances, were phased out. And, in the 1850s, government legislation was introduced to control Indigenous lands and move toward assimilation.[41]

Louis Sawantanan's son, who was also Marguerite Vincent's half-brother, was Grand Chief Nicolas Vincent Tsa8enhohi. Grand Chief Nicolas built on his family's political and diplomatic traditions and was a strong voice in international diplomacy, leading the community through the challenges of the early nineteenth century. He engaged with the British colonial government on behalf of the Wendat community; in 1825, along with three other Wendat chiefs, he visited England to make representations to King George IV, raising issues of land rights and historic agreements between the British and Wendat nations.[42] In 1829, he spoke before the Quebec government, the House of Lower Assembly, outlining the economic difficulties facing the Wendat, many as a result of settler encroachments on Wendat land.[43] This was the environment and community in which Marguerite Vincent was raised: in a politically engaged family, familiar with European social structures and taste; in a culture rich in visual arts with strong traditions of trade and diplomacy; in a belief in the importance of community and a sense of the Wendat as a distinct nation at the core.

The economic and political challenges of the early nineteenth century prompted Wendat development of the existing small-scale souvenir production and, during the course of the century, the range of souvenir works expanded. As well as the hide pouches and moccasins, this included moosehair-embroidered bark-work in domestic objects such as baskets, cigar cases and wall plaques, embroidered with naturalistic floral motifs, such as those seen in the Elgin trays (see Figure 2.1), as well as figural motifs, vignettes of Wendat men and women engaged in daily activities in woodlands scenes.

Marguerite Vincent was well known for her skill in moosehair embroidery and, with her husband and business partner, Paul Picard, she led the development of the Wendat souvenir arts industry in her community and directed the manufacture of the souvenir works. Reflecting the Wendat tradition of complementary work roles, her husband, a noted businessman and hunter, directed the men who provided the raw materials, managed the snowshoe manufacture, and negotiated sales outside Wendake. Men's hunting provided the raw materials to make the snowshoes and souvenir wares made by women, and men also made the snowshoes, sleds and canoes.[44]

Continuing the Wendat business tradition, sales expanded into the wholesale market, with connections to merchants in Montreal, Toronto, Kingston, Quebec, New York and Niagara Falls.[45] According to a guidebook of the period, two stores in Niagara Falls were known to have a 'constant supply of their choicest manufactures ... the greatest market for splendid Indian work'.[46] The discerning traveller or tourist knew to visit Niagara Falls for works of the highest quality and several of the richest Wendat collections now in European museums were acquired there. By the mid-nineteenth century, the community was thriving.[47]

With the English takeover of Quebec, the pilgrimage visits largely ended; however, the village remained a popular destination for Europeans for diplomatic events and to purchase souvenir wares. Regular diplomatic engagements with colonial administrators and visiting dignitaries were held at Wendake, and Wendat chiefs paid official visits to the Governors General at their residence, Château Saint-Louis. Through these formal meetings the Wendat chiefs continued to represent their community politically, maintaining recognition of Wendat cultural identity and sovereignty while enhancing commercial ties and thus assuring a strategic position for their community in relation to Quebec's colonial framework.[48]

These diplomatic events were based in Wendat and Great Lakes traditions of adoption, which dated from the time of Champlain and earlier. The ceremony of honorary adoption invited colonial dignitaries or men who had been of particular assistance to the Wendat to become chiefs.[49] The awarding of a Wendat *nom de guerre* to military officers was another regular event, also based in the adoption ceremony. Members of the British military, their families and broader public were further connected to the Wendat through widely accepted invitations to attend ceremonies in which Wendat leaders were named as war chiefs; thereby the settler community participated in the Wendat community's traditional system of government.[50] People from the European community in Quebec, as well as most members of the Wendat community, attended these events, which included feasting and the exchange of gifts.[51]

The intensely Wendat character of these diplomatic events – the songs, dances and the chiefs' striking appearance in moosehair-embroidered coats, leggings and moccasins, wearing feather headdresses, silver armbands and silver medals – foregrounded Wendat history as military allies and would have enhanced European interest in the souvenir craftwork. Similar in materials, techniques and design to ceremonial works, the souvenir craftwork sold at homes in the community and at markets and stores in Quebec and other cities was an extension of a visual arts tradition of works made for use within the community. This association with Wendat ceremonies would have given the souvenirs added value as

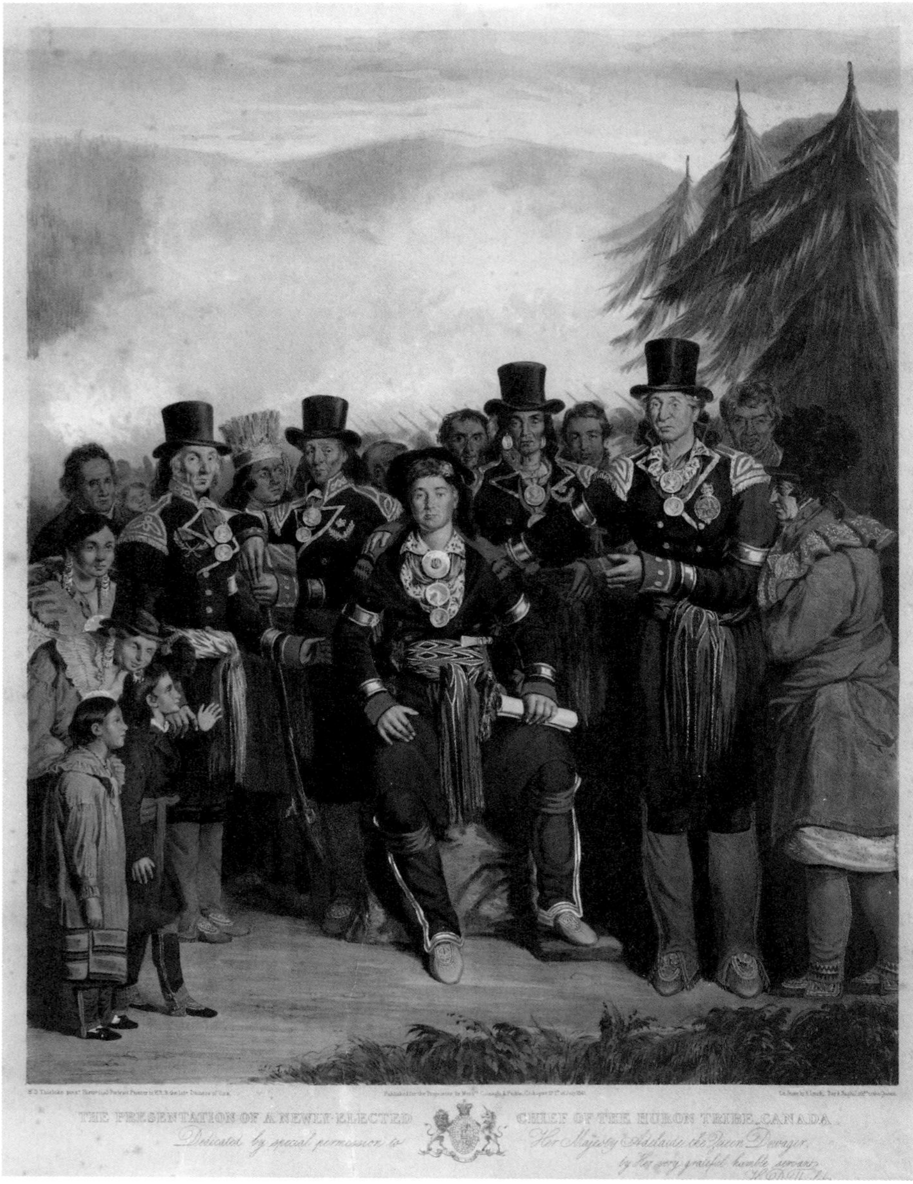

2.5 Henry Daniel Thielcke, *The Presentation of a Newly-Elected Chief of the Huron Tribe, Canada*, 1841.

mementos associated with the Wendat community, appealing to the romantic interest in the exotic in nineteenth-century Quebec and broader settler society.[52] These events and the sale of souvenir arts in Wendake invited European visitors into Wendat territory and encouraged participation in Wendat traditional life. This helped to build relations with settler communities and at the same time asserted their continuing Indigenous culture through the traditional character of the diplomatic ceremonies and visual arts.

Paintings of the period illustrate these ceremonies and the importance of Wendat dress in representing a distinct Wendat identity. A tableau painting by Henry Daniel Thielcke, *The Presentation of a Newly-Elected Chief of the Huron Tribe, Canada* (1841) provides a visual record of this ceremonial complex as well as another Wendat innovation, the memorializing of diplomatic events in text gift-documents (Figure 2.5).

It depicts the honorary adoption of Judge Robert Symes and shows the Judge and members of the community wearing traditional Wendat regalia and clothing, such as moosehair-embroidered epaulettes and moccasins. Judge Symes holds in his hand a bark scroll, a document that was given to honorary chiefs and dignitaries to commemorate formal orations. These bark scrolls were often decorated with moosehair-embroidered motifs and, in one example, with shell wampum beads.[53] They memorialized the ceremonial orations by recording them in French and Wendat text – they became a souvenir of the event and were given to the person to whom the oration had been addressed. In recording diplomatic events for public memory through text, the bark scrolls continued the mnemonic function played by the wampum belts that traditionally accompanied Great Lakes Indigeneous ceremonial oration.[54] The wampum belts were souvenirs, objects that facilitated memory; in the same way that bark scrolls memorialized diplomatic events, commercial bark works memorialized a visit to the Wendat community and, later, notions of the vanishing Indian.

The series of self-portraits by nineteenth-century Wendat artist, Zacharie Vincent,[55] in which he depicts himself in traditional Wendat apparel, further demonstrates the use of traditional dress to publicly represent Wendat identity, cultural heritage and way of life to the European community. This was a public representation of the aesthetic–cultural complex of which the souvenirs were part.[56] In these self-portraits, Vincent usually included a ceremonial calumet pipe, embroidered feather headdress, silver medallion or couette worn by Wendat men and women, a Queen Victoria medal, wampum belt and silver armbands. These objects have symbolic value: Vincent associated them with his public self-representation as a Wendat chief.[57] Vincent's self-portraits were sold to government representatives such as Lord Durham, Lord Elgin, Lord Monk, Princess Louise and others, thereby adding to public awareness of the Wendat aesthetic–cultural community.[58]

Lord and Lady Elgin received what came to be called the Elgin trays, exquisite examples of the Wendat craft tradition, from the Wendake community within the framework of diplomatic events that took place between Wendat and British dignitaries (see Figure 2.1). In January 1852, Chief François-Xavier Tahourenché, Marguerite Vincent's son, led a deputation of chiefs to present a New Year's oration to Lord Elgin at Château Saint-Louis. Six months later, in June 1852, Lord and Lady Elgin made a reciprocal visit to the home of Marguerite Vincent and Chief François-Xavier Tahourenché in Wendake. It is within this diplomatic context of exchanged visits that the trays were probably gifted.

The motifs on the trays symbolically represent the interests of both the Wendat and the Elgins: the initials of Lord and Lady Elgin and heraldic symbols that reference the Elgin and Durham coats of arms mingle with clusters of strawberries, which represent Wendat traditional beliefs. These beautifully worked motifs strongly suggest that the trays were diplomatic gifts rather than commercial souvenirs. European heraldic devices were again brought together with Wendat traditional motifs in the bark scroll given to Folke Cronholm in 1905 as a diplomatic gift during his honorary adoption.[59]

The visits between the Vincent and Elgin families, the diplomatic context of the trays, and the technical perfection combined with virtuosity of design and formal arrangement have led to assigning attribution to Marguerite Vincent La8inonke. However, community knowledge of the history of the mid-nineteenth-century processes of moosehair-embroidery production suggests that the creator of the trays could also have been one of the artist–needlewomen who produced works for the Vincent–Picard family souvenir enterprise.[60] Although attribution, in this current research, remains uncertain, stylistic similarities between the Elgin trays and two other mid-nineteenth-century works suggest a common oeuvre. Stylistic analysis of the Elgin trays finds points of similarity with two other works that may be by the same hand. First, a large blue flower, possibly a convolvulus, with distinctive black outlining the centre, appears in one of the Elgin trays, in a bark wallet in the collection of the Canadian Museum of Civilization, and in a tray from the Elgin collection now at the Fenimore Art Museum (Cooperstown, New York).[61] All three works demonstrate a fine, light style in the individual motifs, which are integrated into complex arrangements to create a harmonious overall design. Characteristic of these works is a high degree of naturalism in form as well as three-dimensionality achieved through the use of graduated colours and shading in the moosehair stitches. This approach, through archival research and stylistic analysis, may offer further opportunities to associate bodies of work with specific artists. It may be that artists and their oeuvres can be known by their stylistic signatures rather than by their names; this identification is an achievement, bringing recognition to individual women.

Wendat success in nineteenth-century trade attests to the success of the programmes of diplomatic events and supported the burgeoning souvenir sales as well as military contracts. Increasing numbers of European visitors were coming to Quebec in this period, resulting in growing markets for souvenirs.[62] By 1870, approximately 60 of the total 76 families in the community made their living from the production of embroidered moccasins and bark work as well as snowshoe moccasins. Newspaper accounts of the period recorded the volume of embroidered souvenirs sold, and the journal of Wendat chief François-Xavier Picard Tahouenché records the large scale of the military contracts.[63] A contract with the British government dated 28 September 1837 records a sale of '1,000 paires de souliers mou, et 400 paires de raquettes apres cela encore 344 paires de mesme souliers et 336 paire de raquettes' (1,000 pairs of soft shoes and 400 pairs of showshoes, after that a further 344 pairs of the same shoes and 336 pairs of snowshoes).[64] The community was thriving and newspaper articles of the period described the elegant homes of the community: 'l'abondance se répandit dans le village' (there was abundance in the village).[65]

The tradition of military support was integral to Wendat alliances with the French and then the English, supported by ceremonies that renewed the closeness between the communities. The Wendat fought beside the British during the American War of Independence and the War of 1812.[66] In 1838, during a period of British–American political tension and the anticipation of American attack, the Wendat remained closely connected with the British garrison at Quebec.[67] The Wendat community maintained its military commitments to support the British from the late eighteenth through the nineteenth centuries, although political tensions arising from the possibility of military engagements in the Canadas largely ceased after 1840.

Complicating the Contact Zone

Quebec and the region around it was the site of these historical relations between the Wendat and European settlers – as trade partners, military allies and neighbours. The region can be understood as a contact zone, the term conceptualized by Mary Louise Pratt to describe a 'space of colonial encounters … in which peoples geographically and historically separated come into contact with each other and establish ongoing relations'.[68] However, my discussion of the Wendat souvenir craft and its situation within a complex of Wendat cultural–aesthetic practices complicates Pratt's description of relations between peoples in this encounter and adds fresh perspectives. Pratt describes these encounters in terms of 'colonized subjects' in relations with colonial governments characterized by 'coercion, radical inequality, and intractable conflict'.[69] However, the narrative of the Wendat souvenir craft and the community's relations with European nations demonstrates that a contact zone could also be a site for the assertion of Indigenous autonomy, agency and distinct identity, while at the same time building harmonious relations between communities. There may have been coercion, radical inequality in power relations and intractable conflict, but Indigenous nations did not necessarily act as subservient colonized subjects, nor necessarily see themselves in that way.

The history of the Wendat nation, explored through the lens of their souvenir craft tradition, demonstrates the strength and unity of their community and the continuity of their values and worldview, expressed in their practices of diplomacy, trade and military alliance. These practices demonstrate the power the Wendat claimed over decisions regarding their future as a community, through their traditional practices that over centuries successfully engaged with colonizing forces. The key point is that these diplomatic practices were assertions of Wendat sovereignty: the Wendat continued their position as allies, not subjects. When British policies toward Indigenous peoples after the War of 1812 changed, moving from a period of conflict to an age of settlement and business expansion, colonial governments wanted Indigenous communities' land and Indigenous peoples to become assimilated subjects, not military allies who described themselves in the terms of sovereign nations. However, the continuity of Wendat diplomatic traditions demonstrates a persistent autonomy in trade and military allegiance, sustained through strategic initiatives in these spheres. Diplomatic ceremonies not only asserted Wendat identity and cultural traditions and affirmed trade relations; they were also serious undertakings, commitments to the British that Wendat warriors would fight beside them in time of war, as allies, not subjects. Reflecting their history of diplomatic skills, Wendat chiefs asserted their sovereign identity with tact, addressing issues of identity with dignity and persuasion. As this history has demonstrated, visual arts, among them the souvenir craft tradition, were central to these initiatives.

Caroline Gros-Louis

Toward the end of the 1870s, the moosehair-embroidery production in which the Vincent and Picard families had played a central role moved to the Gros-Louis and Sioui families.[70] The women of the Vincent, Picard, Gros-Louis and Sioui families who produced Wendat craftwork were part of a community of artists whose specialized knowledge was passed on through the generations. The Wendat virtuosity in their craft tradition continued, as did its commercial success and its role in diplomacy. Caroline Gros-Louis, who was recognized for the artistry of her moosehair-embroidered souvenir and ceremonial artworks, was taught the techniques and traditional aesthetic of Wendat craft by her mother, Marie Robigaud (Mrs Etienne Gros-Louis, 1831–1921), herself of the era of craft production of Marguerite Vincent La8inonke.[71]

In the early twentieth century, a period when the names of Indigenous artists were more likely to be recorded by collectors, Caroline Gros-Louis's works entered two significant collections: the Canadian Museum of Civilization acquired commercial commissions, and the collection of Swedish diplomat Folke Cronholm received diplomatic gifts. One of her best-known works is the chief's outfit embellished in exquisite moosehair embroidery presented to Cronholm as a gift in an adoption ceremony in 1905, illustrating both the continuity of the moosehair-embroidery craft and the diplomatic ceremony of which it was an integral part (see Figure 2.2). This outfit was displayed in the window of the Holt Renfrew store in Quebec City in 1906; a newspaper article of the period named Caroline Gros-Louis as the artist and described the embroidery as 'a work of art'.[72] Caroline Gros-Louis continued the tradition of stylized floral imagery seen in mid-nineteenth-century work such as that found in the Bata Shoe Museum moccasins (see Figure 2.4). At the same time, the work reflects Gros-Louis's own style, with more simplified designs, bold colours and focus on a sinuous line in harmonious, balanced arrangements characteristic of the Wendat aesthetic. Caroline's work in the chief's outfit also continued Wendat diplomatic traditions. This gift, a central symbol of the adoption ceremony, formalized diplomatic relations between the Wendat and the Swedish consul, the representative of the Swedish king; this continued Wendat practices dating from the early seventeenth century and before.

In conclusion, study of the Wendat souvenir craftwork demonstrates their community's strong sense of identity as a nation, supported by the artistic community of generations of Wendat women. Over the centuries there were changes in visual arts traditions, innovations in design and style, to reflect new markets and tastes as the community deftly adapted its diplomatic and creative traditions to the shifting circumstances of the European colonial presence. At the same time, the essence of the craft-art continued, evident in the materials and the virtuosity of technique and design, creating beautiful objects for ceremony and trade. These craft-arts moved across time and space as they sustained Wendat culture and identity.

Primary Sources Cited

Alexander, James Du Pre, Third Earl of Caledon, Journals and correspondence. 1838–1842, D2433, Public Record Office of Northern Ireland (PRONI), Belfast.

Fonds, Ethnology Division, Canadian Museum of Civilisation, Gatineau, Box 61, Folder 5.

Graves Simcoe, Mrs John, *The Diary of Mrs. John Graves Simcoe* (Toronto: William Briggs, 1911).

Picard Tahourenché, François-Xavier, Journal de François-Xavier Picard Tahourenché,

Archives de Conseil de Wendake, Quebec.

Secondary Sources Cited

Balvay, Arnaud, *L'Épée et la Plume* (Québec: Les Presses de l'Université Laval, 2006).

Beaulieu, Alain, 'Les Hurons et la Conquête', *Recherches Amérindiennes au Québec*, 30/3 (2000), pp. 53–63.

Bouchard, Marie, 'Curating in Relation to Community', in *Making a Noise!* (Banff: The Banff Centre for Continuing Education, 2004), pp. 206–21.

Champlain, Samuel de, *The Works of Samuel de Champlain*, 3 vols (Toronto: The Champlain Society, 1922–36).

Charlevoix, Père, de la Compagnie de Jésus, *Journal d'un Voyage fait par l'ordre du roi* (Paris: Rollin fils, 1744).

Delâge, Denys, 'La tradition de commerce chez les Hurons de Lorette-Wendake', 30/3 (2000), pp. 35–51.

Druin, Pierre and Margot Rankin, based on work by Jean-Pierre Sawaya, *Marguerite Vincent 'Lawinonkié'*, Formulaire de Demande-Personne, 2000–42 (Ottawa: Historic Sites and Monuments Board of Canada, 2000).

Every Man His Own Guide to the Falls of Niagara (Buffalo: Paxon & Stevens, 1846).

Frayling, Christopher and Amanda Game, 'What is Craft?' from the Victoria & Albert Museum. Available at: http://www.vam.ac.uk/content/articles/w/what-is-craft [accessed 30 November 2011].

Haefeli, Evan and Kevin Sweeney, *Captors and Captives* (Amherst and Boston, MA: University of Massachusetts Press, 2003).

Labelle, Marie-Dominique and Sylvie Thivierge, 'Un peintre huron du XIXe siècle: Zacharie Vincent', *Recherches Amérindiennes au Québec*, 11/4 (1981), pp. 325–33.

LeMoine, J.M., *Picturesque Quebec* (Montreal: Dawson, 1882).

Lindsay, L'Abbé Lionel Saint-George, *Notre-Dame de la Jeune Lorette* (Montreal: La Cie de Publication de la Revue Canadienne, 1900).

Long, John, *The Voyages and Travels of an Indian Interpreter and Trader* ([J. Long, 1791], Toronto: Coles Pub. Co., 1971).

Lotman, Yuri, *Universe of the Mind* (London: Tauris, 1990).

Phillips, Ruth, 'A Proper Place for Art or the Proper Arts of Place?', in Lynda Jessup with Shannon Bagg (eds), *On Aboriginal Representation in the Gallery* (Ottawa: Canadian Museum of Civilization, 2002), pp. 45–71.

Pratt, Mary Louis, *Imperial Eyes* (London and New York: Routledge, 1992).

Sawaya, Jean-Pierre, 'Les Amérindiens domiciliés et le protestantisme au XVIIIe siècle: Eleazar Wheelock et le Dartmouth College', *Historical Studies in Education/revue d'histoire de l'éducation* (Fall/automne 2010), pp. 1–21.

Sioui, Anne-Marie, 'Un oeuvre engagé?', *Recherches Amérindiennes du Québec*, 11/4 (1981), pp. 335–7.

Steckley, John, *A Huron–English/English–Huron Dictionary* (Queenston: The Edwin Mellen Press, 2007).

Tanguay, Jean, *Margurite Vincent La8inonkie*, Rapport Supplementaire 2007–29 (Ottawa: Historic Sites and Monuments Board of Canada, 2007).

Tooker, Elizabeth, *An Ethnography of the Huron Indians 1615–1649* (Midland.: The Huronia Historical Development Council, 1967).

Trigger, Bruce G., *The Children of Aataentsic* (Kingston and Montreal: McGill-Queen's University Press, [1976] 1987).

'Uniforme de Chef Huron', *Le Soleil*, 7 June 1906.

Vincent Tehariolina, Marguerite, *La Nation Huronne* (Quebec: Éditions du Pélican, 1984).

Whitehead, Ruth, *Micmac Quillwork* (Halifax: Nova Scotia Museum, 1982).

Notes

1 'Huron' was the name given to the people of the Wendat confederacy by the French, 'Wendat' is the name of the community in their language.

2 Ruth Phillips, 'A Proper Place for Art or the Proper Arts of Place?', in Lynda Jessup with Shannon Bagg (eds), *On Aboriginal Representation in the Gallery* (Gatineau, Quebec, 2002), p. 63.

3 Christopher Frayling and Amanda Game, 'What is Craft?' in the Victoria & Albert Museum. Available at: http://www.vam.ac.uk/content/articles/w/what-is-craft [accessed 30 November 2011].

4 Marie Bouchard, 'Curating in Relation to Community', in Lee-Ann Martin (ed.), *Making a Noise!* (Banff, 2004), p. 219.

5 The use of '8' in the Wendat language indicates the pronunciation of what is roughly a 'whu' sound. In *A Huron–English/English–Huron Dictionary*, John Steckley describes it as a 'u' written over an 'o' (Queenston: The Edwin Mellen Press, 2007), p. 10.

6 Communication with Mireille Siouï, 26 June 2012.

7 Jean Tanguay, *Marguerite Vincent La8inonkie, 'La Femme Habile aux Travaux d'Aiguille'*, Rapport Supplémentaire 2007–29 (Ottawa, 2007), p. 1.

8 Mireille Siouï in discussion, Congrès d'Études Wendat et Wyandot, 14 June 2012. The Vincent–Picard family also had works made on commission; hence it is possible that the trays are by another woman artist from the Wendat community.

9 Mireille Siouï in discussion, Congrès d'Études Wendat et Wyandot, 14 June 2012.

10 Email communication with Mireille Siouï, 21 June 2012. She stated 'en effectuant ces travaux à la maison, les femmes huronnes ont contribué à la transmission des savoir-faire traditionnels'. My translation.

11 Ruth Whitehead, *Micmac Quillwork* (Halifax, 1982), p. 30.

12 Ibid.

13 Mrs John Graves Simcoe, *The Diary of Mrs. John Graves Simcoe* (Toronto, 1911), p. 67.

14 Bruce G. Trigger, *The Children of Aataentsic* (Montreal and Kingston, [1976] 1987), p. 364.

15 Ibid.

16 Samuel de Champlain, *The Works of Samuel de Champlain* (3 vols, Toronto, 1922–36), vol. 2, pp. 68, 80, 128.

17 Ibid., vol. 2, pp. 99–101, Champlain's valour in battle; pp. 104–5, Wendat invitation: '[ils] me dirent si ie ne desirous pas aller en leur pays pour les assister tousiours comme freres. Ie leur promis' ([they] asked me if I would like to go with them to their country and help them always as brothers. I promised I would).

18 Champlain, *The Works of Samuel de Champlain*, vol. 2, pp. 109–10.

19 Ibid., giving woven belt to Henri IV; ibid., vol. 3, pp. 132–3, description of decorative work by Wendat and Algonquin women artists.

20 Ibid., vol. 2, pp. 69, 102, 106, 159–60, diplomatic gifts.

21 John Long, *The Voyages and Travels of an Indian Interpreter and Trader* ([J.Long, 1791] Toronto, 1971), pp. 44–9.

22 Champlain, *The Works of Samuel de Champlain*, vol. 2, pp. 121–2.

23 Ibid., vol. 3, pp. 47, 133–4.

24 Arnaud Balvay, *L'Épée et la Plume* (Québec, 2006), pp. 216, 221, 229; Dépenses générales, 1747, Reel F118: MG1, Serie C11A, vol. 117, Fol. 95–116 (Ottawa).

25 Canadian Museum of Civilization: III-H-426, a moosehair-embroidered pouch collected by German soldiers of the Duke of Gotha either in the period of the American Revolutionary War in 1776 or in the first decades of the nineteenth century; III-H-429, III-H-430, III-H-431, a family of dolls, labelled 'Huron', collected in 1788.

26 Trigger, *The Children of Aataentsic*, pp. 34–45, 100.

27 Elizabeth Tooker, *An Ethnography of the Huron Indians 1615–1649* (Midland, Ontario, 1967), p. 59, Wendat women's preparation of hides.

28 *Journal of François-Xavier Picard Tahourenché*, 18 mai 1849. Archives, Wendat Band Council, Wendake, Quebec.

29 L'Abbé Lionel Saint-George Lindsay, *Notre-Dame de la Jeune Lorette* (Montreal, 1900), pp. 133, 140.

30 Charlevoix, *Journal d'un Voyage*, pp. 83–4.

31 Ibid.

32 Evan Haefeli and Kevin Sweeney, *Captors and Captives* (Amherst and Boston, MA, 2003), pp. 2, 213–15.

33 Alain Beaulieu, 'Les Hurons et la Conquête', *Recherches Amérindiennes au Québec*, 30/3 (2000), pp. 56–7.

34 Ibid., p. 60.

35 Jean-Paul Sawaya, 'Les Amérindiens domiciliés et le protestantisme au XVIIIe siècle', *Historical Studies in Education/Revue d'histoire de l'éducation* (Fall/automne 2010), p. 11.

36 Ibid., p. 12.

37 Ibid., pp. 5–6.

38 Ibid., p. 15.

39 Ibid., pp. 15–16.

40 Tanguay, *Margurite Vincent La8inonkie*, p. 13.

41 1850: 'An Act for the Better Protection of the Lands and Property of Indians in Lower Canada'; 1857: 'Act for the Gradual Civilization of the Indian Tribes in the Canadas (Gradual Civilisation Act)'; this act introduced the principal of legal emancipation and made assimilation its goal. Enfranchisement made the Indigenous applicant a British subject but would end his and his family's Aboriginal status and rights; *Copies of Extracts of Correspondence Since 1st April 1835 Between the Secretary of State for the Colonies and the Governors of the British North American Provinces respecting the Indians in those Provinces* (London, 1839), p. 2, recommendations that gifts given to Indigenous nations as allies be phased out.

42 Marguerite Vincent Tehariolina, *La Nation Huronne* (Quebec, 1984), p. 82.

43 Ibid., p. 10.

44 Ibid., pp. 2–3, 12.

45 Ibid., p. 14.

46 *Every Man His Own Guide to the Falls of Niagara* (Buffalo, 1846), pp. 69–70.

47 Pierre Druin and Margot Rankin, based on work by Jean-Pierre Sawaya, *Marguerite Vincent 'Lawinonkié'*, Formulaire de Demande-Personne, 2000–2042 (Ottawa, 2000), p. 2159.

48 Vincent Tehariolina, *La Nation Huronne*, pp. 321–39.

49 J.M. LeMoine, *Picturesque Quebec* (Montreal, 1882), p. 449. Mr W. Cooper, who accompanied Grand Chief Nicolas Vincent Tsa8enhohi to England in 1824, was made an honorary chief, known as Chief Tourhaunché.

50 Picard Tahourenché, *Journal*, 1837, 1840.

51 Ibid.

52 Marie-Dominique Labelle and Sylvie Thivierge, 'Un peintre huron du XIXe siècle: Zacharie Vincent', *Recherches Amérindiennes au Québec*, 11/4 (1981), p. 326.

53 Bark scroll, Am 1887.1208.19, British Museum, London, England.

54 Yuri Lotman, *Universe of the Mind* (London, 1990), pp. 246–7.

55 Zacharie Vincent was not directly related to Marguerite Vincent La8inonkie.

56 Labelle and Thivierge, 'Un peintre huron', pp. 325–9.

57 Anne-Marie Sioui, 'Un oeuvre engagé?', *Recherches Amérindiennes du Québec*, 11/4 (1981), p. 335.

58 Ibid.

59 Folke Cronholm collection, 1936.22.0017, Etnografiska Museet, Stockholm, Sweden.

60 Mireille Siouï, 14 June 2012.

61 Wendat wallet, III-H-491, Canadian Museum of Civilization, Gatineau, Quebec; Wendat tray, T283, Fenimore Art Museum, Cooperstown, New York.

62 James Du Pre Alexander, Third Earl of Caledon 1840: B/8/22.

63 Tanguay, *Margurite Vincent La8inonkie*, p. 14. My translation.

64 Picard Tahourenché, *Journal*, Contrat du Gouvernement, 28 septembre 1837. My translation.

65 Denys Delâge, 'La tradition de commerce chez les Hurons de Lorette-Wendake', *Recherches Amérindiennes du Québec*, 30/3 (2000), p. 46.

66 LeMoine, *Picturesque Quebec*, pp. 457, 464–5.

67 Alexander, 1838: B/8/6.

68 Mary Louis Pratt, *Imperial Eyes* (London and New York, 1992), p. 6.

69 Ibid., pp. 7, 6.

70 Druin and Rankin, with Jean-Pierre Sawaya, *Marguerite Vincent 'Lawinonkié'*, p. 2160.

71 III-H-2, List of Culture Area III works. Ethnological Division, pt. 2, Box 61, Folder 5. Fonds of the Ethnological Division, Canadian Museum of Civilization, Gatineau, Quebec.

72 'Uniforme de Chef Huron', *Le Soleil*, 7 June 1906.

3

'To foster and encourage the study and practice of Māori arts and crafts': Indigenous Material Culture, Colonial Arts and Crafts and New Zealand Museums

Conal McCarthy

Introduction

The image reproduced in Figure 3.1 shows a plate decorated with a picture of a Māori woman and child, a native Madonna, surrounded by a band with a curvilinear Māori pattern known as *kōwhaiwhai*. Within that pattern the Māori word '*kiaora*', meaning hello, forms a link with stylized tattooed faces. It is a good example of the way in which the Arts and Crafts movement in New Zealand adopted Māori motifs to create a distinctive national style. Such objects have an ambiguous status in the present: collectable, yet also seen as kitsch, and sometimes condemned as culturally insensitive, particularly because they transgress cultural prescriptions (keeping the *tapu* or sacred 'head' separate from the 'profane' food) and generally because indigenous imagery is taken without permission and used to add something Polynesian to the construction of a European settler imaginary. Critiques of the western use of non-western culture are very common in the literature of museum studies, art history and postcolonial studies, in which indigenous material culture is said to have been invented or appropriated during colonization by Europeans.[1]

This chapter challenges that position, showing that the colonial encounter was productive as well as destructive of Māori material culture in Aotearoa New Zealand. In doing so it draws on a recent revival of interest in craft, now being reassessed through interdisciplinary research in fields such as material culture studies, with a new appreciation for both the materiality and relationality of craft. In this volume, scholars examine craft as material culture in settings around the world, exploring connections between community, place and politics. In this spirit, I argue that the abundant evidence

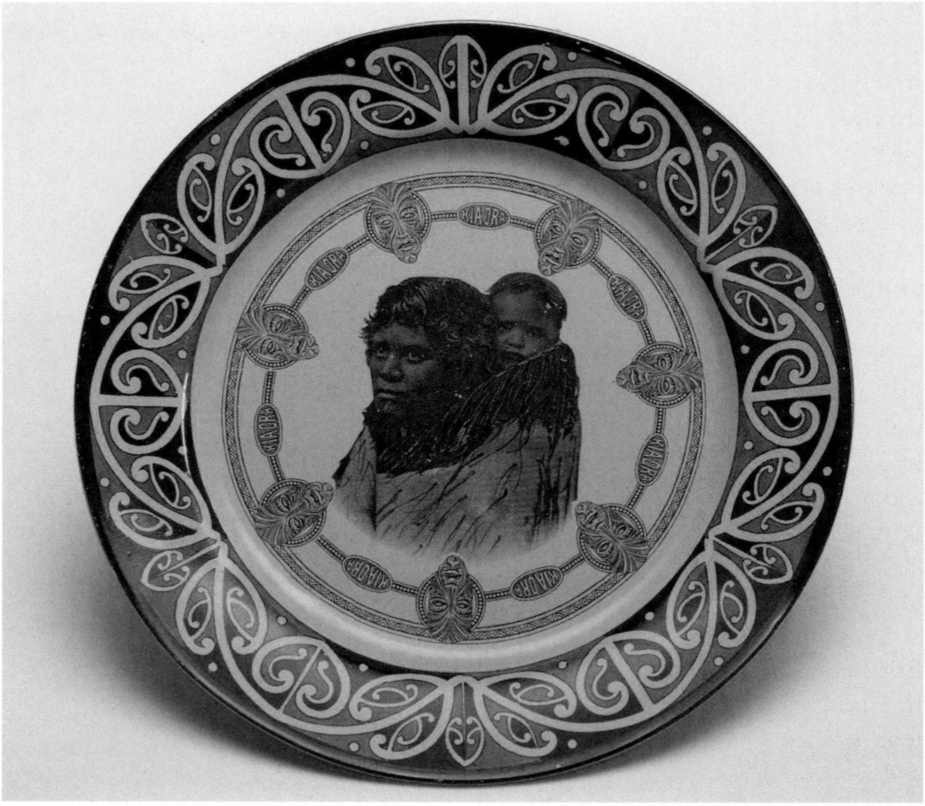

3.1 Royal Doulton plate with Māori designs showing arts and crafts influences. The Māori word '*Kiaora*' means hello. Bone-china cabinet plate with printed decoration and hand-painted detailing, c. 1925. Overall dimension (height) 25 mm.

of Māori agency in facilitating the widespread interest in native designs in New Zealand shows how indigenous people participated in a colonial culture of art and craft. This story is a transnational one, found in several settler colonies. Ruth Phillips describes the intersection of native craft and colonial culture as 'transcultural aesthetic processes' that are articulated through the asymmetrical power relations of colonial societies, not always to the complete and inevitable advantage of colonizer over colonized.[2]

How did this 'discovery' of Māori arts and crafts in New Zealand come about? What role did Māori themselves play in the process, and how did it serve their own interests? In tracing the ways in which Māori responded to opportunities, I chart a rise in the status of indigenous decorative arts among the dominant Pakeha (European New Zealanders), seen in the positive reception of Māori objects in museum exhibitions and the adoption of Māori elements into public art and architecture. The development of the School of Māori Arts and Crafts also reveals how government preservation initiatives were steered by individuals such as the prominent Māori politician Āpirana Ngata. While working within the system, he resisted the dominant ethos of assimilation and used arts and crafts as a political tool, with some success, in the struggle to advance Māori causes.

Art, Craft, Indigenous Culture, Settler Colonialism

In the nineteenth to first half of the twentieth century New Zealand saw itself as the Britain of the South Seas in a vision of a modern European nation that, as one politician put it, incorporated a dash of 'the finest native race in the world'.[3] Māori–Pakeha relations in the interwar period were dominated by an official policy of assimilation, with Māori struggling to create and maintain their own spaces in government, tribal economies, community affairs, sport and the arts. Like other native peoples in settler colonies such as Australia and Canada, Māori people negotiated with modernity by engaging with state-sanctioned projects of race uplift, preservation and commemoration, attempting to bend these plans to their own purpose.[4] The leaders of the Young Māori Party, parliamentary politicians such as Āpirana Ngata, Peter Buck and Māui Pōmare, along with community leaders such as Te Puea Herangi of the Tainui tribes in the Waikato region, managed to wrest concessions from the majority culture and toiled to preserve and maintain their language and customs in the face of considerable odds. While this cultural heritage was based on a revival of the past, paradoxically they saw it as forward-looking, demonstrating their survival in the present and demanding a stake in the future of the modern nation.[5]

Since the colony of New Zealand was regarded as a loyal outpost of Empire, more British than the British, it is not surprising that the Arts and Crafts movement was transplanted to the South Pacific by the end of the nineteenth century. Over the next four decades its influence was felt in architecture, art education, handicrafts, textile and interior design.[6] This can be seen in the New Zealand International Exhibition in Christchurch in 1906–7, where the British Arts and Crafts section, a highlight of the attractions, displayed objects selected by Walter Crane, President of the Arts and Crafts Exhibition Society in London.[7] At the same time Māori motifs were often used by New Zealand artists looking for a convenient vernacular source for their applied arts to distinguish them from other British colonial artists. As Ann Calhoun points out, this may be interpreted today as a form of appropriation, but other scholars, such as anthropologist Nicholas Thomas, stress the hybrid and contested nature of indigenous art within colonial culture. He sympathetically explores the borrowing of Māori and Aboriginal designs within Australian and New Zealand settler nationalism, recognizing the settler/native dialogue that was never one-way or one-sided.[8]

The Arts and Crafts movement began in England as a reaction against industrialization and a return to the aesthetic values of craftsmanship; but by the late nineteenth century it had grown and spread, sometimes providing a forum through which political categories of the time were challenged. At times it appears to have been quite egalitarian, its philosophy of natural materials and traditional handicrafts evincing a more democratic openness, reflecting the socialist principles of William Morris and other adherents, with the result that we arguably see greater involvement by women, working-class artisans and even subalterns than in the rarefied domain of fine art. The English Arts and Crafts movement had a major impact on the colonial art world of New Zealand, and contributed to the shifting aesthetic appreciation that saw

Māori carving ascend to the status of art/craft, a process explored in the next section. Despite the domination of European categories and practices, Māori objects did enter the colonial canon of art and craft through their inclusion in museums, exhibitions and other media in New Zealand. Māori objects shifted from curio, to specimen, and then artefact, and the things once considered primitive were by 1900 praised variously as 'native art workmanship', 'art handicraft' and 'arts and crafts'.[9]

How did this transformation come about? According to James Belich's revisionist historical framework, indigenous people in this settler colony fared somewhat better than in other parts of the British Empire as a result of the way native culture, flora and fauna became part of the iconography of 'Māoriland'. With European colonization Māori people suffered as their land was appropriated in much the same way as in other British territories. However, during the process of 'recolonization' from the 1880s, a complex process in which New Zealand was tied even more closely to the mother country (or *recolonized*), the people formerly regarded as 'savages' were incorporated into the 'better Britain' of the South Pacific, their material culture reconstituted as almost-art, befitting Māori status within the white settler colony as almost-white. So when Māori became 'whitened' as 'brown Britons', their 'arts and crafts' were assimilated into a nascent form of colonial identity for the settler nation, a romantic vision of 'Māoriland' constructed from native plants, animals and elements of Māori visual and material culture.[10]

The objects that Pākehā began to call 'art' were not actually made or perceived *as art* (in the western sense) by their original creators and audiences. There was no word for 'art' in the Māori language. Polynesian societies did not have a concept of beauty or aesthetic form corresponding to western ideas. The Māori word *taonga* (treasure) was not a synonym for art, despite its popular usage since the 1980s to refer to museum artefacts. Māori words currently used in English for art, such as *toi* or *whakairo*, were previously adjectives rather than nouns. *Whakairo*, for example, denoted design or pattern and not an object itself – it was not carving but *anything* that was decorated, such as weaving, clothing and so on.[11] A *whare whakairo* or carved meeting house was and is regarded as literally the body of an ancestor, or *whare tupuna*. The tribe gathers inside the belly of their founding ancestor to make important decisions in the presence of carved figures around the walls regarded by descendants as living embodiments of individual ancestors. The 'arts' of the meeting house – *whakairo* or wood carving, *tukutuku* or lattice work, and *kōwhaiwhai* or rafter painting – were originally seen as an expression of group identity, not something valued for their own sake. Anthropologist Roger Neich calls them painted and carved histories, and reminds us that carving was not 'isolated and emphasised as a fine art in its own right to the same extent as in the European system of classification'.[12]

But the absence of a pre-European 'art' tradition is not the whole story. Indeed, as Māori were exposed to the western culture of art, they absorbed and adapted it as they did other new concepts and technologies. So by the early twentieth century, alongside continuing customary practices such as weaving and carving, some Māori people were experimenting with new media, materials and forms within European art markets, such as the extraordinary carver Tene Waitere whose experimentation with western naturalism produced remarkable syncretic 'sculptures' in wood.[13] After the Second World War, as Damian Skinner has shown, Māori arts and crafts, by now thought of as 'traditional', was joined by a succession of new artistic movements that each borrowed from western sources.[14] Today 'Māori art' in various forms is largely accepted within the canon of New Zealand art.

This is not to say that Māori did not have their own framework of visual and material culture before contact, which to an extent shared traits of the European art culture. In this aesthetic hierarchy viewers admired and responded to places, objects and performances for their intrinsic qualities of *ihi* (power), *wehi* (awe/fear) and *wana* (authority).[15] The indigenous concept of *taonga* (cultural treasure, property), revived in recent times within Māori museum practice, has been described as 'a special category of objects that instantiate and perpetuate relationships through time'.[16] In contrast to the western separation of subject and object, nature and culture, in the Māori world people and things have close relations that, as Amiria Henare points out, collapse spatial and temporal boundaries. This indigenous ontology expressed through the concept of *taonga* is not unlike recent western theories such as Alfred Gell's notion of art and social agency and Bruno Latour's idea of actor networks that refuses the dichotomy of subject/object, human/non-human.[17] It could be said, then, that Māori 'art' was not discovered, but produced, and Māori were involved in this production because it allowed them to engage with settler society. In so doing they sought not just the preservation of the objects themselves, but the revitalization of practices and traditions connected to them.[18]

'Art Workmanship': The Rising Status of Māori Carving

Māori objects were generally not collected, displayed and discussed together with the European fine arts of painting and sculpture for most of the nineteenth century, until a small group of students, artists, collectors and enthusiasts began to pay more attention to them. One of the first signs of change came in the mid-1880s when Auckland art students began to notice the beautiful things in local museums, bemoaning the fact that they were largely not appreciated by their countrymen. When art students' exhibitions were criticized for declining 'into a sort of Māori curiosity shop', the artist Alfred Sharpe leapt to their defence, responding that 'We do not call them curios; we call them Māori works of art; and they are so in every sense of the word'.[19]

Augustus Hamilton was the most high-profile supporter of the 'new-found' artistic qualities of Māori material culture, and also the director of the Colonial Museum in the capital city of Wellington from 1903 to 1913. An amateur naturalist, ethnologist and avid collector, Hamilton delivered a lecture on Māori art in 1888 while honorary curator of the museum of the Hawkes Bay Philosophical Society in Napier.[20] Along with other Pakeha collectors, Hamilton was instrumental in organizing the large South Seas court at the New Zealand International Exhibition in Dunedin in 1889–90. The extensive display demonstrated the 'artistic' abilities of the Māori 'artist' and included a fully carved house from his own collection that was hailed as a 'magnificent specimen of Māori workmanship'.[21]

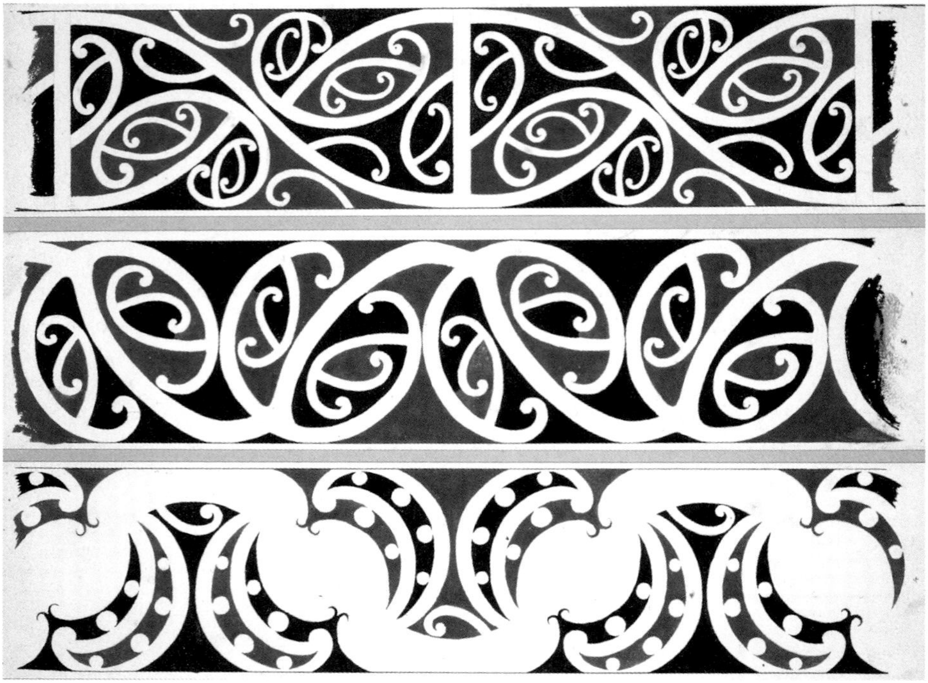

3.2 *Kōwhaiwhai* pattern copied from rafters in a meeting house near Gisborne by Herbert Williams in the 1890s and reproduced in Augustus Hamilton's book *Māori Art* (1901).

After the success of this exhibit, Hamilton researched and published a large illustrated book on the subject that became popularly known as *Māori Art*. The Arts and Crafts aesthetic was reflected in the design of the volume itself, with sinuous Māori motifs embellishing the spine and cover, and a fanciful 'Māori' design on the frontispiece in which a *pare* (lintel) was incorporated into a graceful linear pattern. Figure 3.2 shows one of several *kōwhaiwhai* patterns included in the book, copied from rafter patterns in meeting houses by Archdeacon Herbert Williams. These intriguing curvilinear designs with the now famous '*koru*' or fern-frond motif became widely disseminated, adorning all manner of European objects in the years to come (including the plate seen in Figure 3.1), albeit aestheticized according to European tastes. In his introduction to the first volume, Hamilton wrote that 'Māori art' had reached a very high level among the native people. The voluminous contents of this book, with photographs of decorated objects in wood, stone, bone and ivory, was extremely influential. Serialized and published in several volumes by the New Zealand Institute between 1896 and 1902, it was responsible in large part for inventing the very notion of Māori 'art'.[22]

Hamilton's exploration of Māori 'ornament' reflected artistic fashions in Europe. Scholars have suggested that Hamilton may have been exposed to the work of ethnologists such as A.C. Haddon, who published work on primitive design in the Pacific.[23] Doubtless he knew of trends in display at the Pitt Rivers Museum at Oxford, where an exhibition on 'human forms in savage art' was staged in 1894–95. In 1897 the museum's curator, Henry Balfour, lectured on 'decorative art of the primitive peoples'.[24] Yet the idea that Māori carving could be seen as art was a surprise to many colonists. Author and bibliophile Johannes Anderson wrote to Hamilton telling him:

> Certain it is that the first sight of your book on Māori Art is as a revelation to the great majority of people ... they had no idea the Māori was capable of such beautiful work.[25]

When he became director of the Colonial Museum in 1903, Hamilton was charged with administering the recently passed Antiquities Act stopping the sale of Māori artefacts overseas, and building up the Māori collections for a planned National Māori museum, initiatives championed by James Carroll, the Māori politician who was Minister of Native Affairs. Hamilton began an ambitious acquisition programme and pressed the government to provide a new building to house 'the pick of the finest art of the Māori artist'.[26] In 1904 he delivered a public lecture on Māori 'art workmanship', arguing that the essential elements of 'Māori decorative art' be applied to buildings and utilitarian things in order for it to express 'a national characteristic', as well as being a 'memorial to the race who created and developed it'.[27] It is possible to discern in Hamilton's writing the influence of John Ruskin, William Morris and the English Arts and Crafts movement.[28] In a later paper Hamilton describes Māori art with direct reference to the ideas of Ruskin.[29] The English Arts and Crafts movement sought a return to the natural forms and materials of medieval and folk traditions, evident in the 'hand of the maker' on beautifully crafted objects.

Before the First World War, museum displays reflected the growing admiration for customary Māori carving, the form of indigenous material culture closest to European sculpture. In contrast to earlier displays that mixed up Māori carving with other collections (like natural history), in the 1900s it came to be regarded as a decorative art form in its own right and was consequently displayed by itself, unencumbered by comparison with other *ethnographica*. Nevertheless, though it was considered capable of aesthetic beauty, Māori art-craft such as wood carving was still not considered the equal of fine art (indeed nor was European craft). Instead of being displayed next to European painting or sculpture, Māori carvings merely provided props to be incorporated into the portraits and historical paintings of Pākehā artists interested in Māori subjects. These aesthetic categories were still impervious, with separate buildings, spaces and distinctive display conventions.[30] But the politics of indigenous crafts thereafter took new forms in New Zealand.

The New Zealand Exhibition in Christchurch of 1906–7, with which Hamilton was closely involved, is a good example of how Māori actually engaged with the notions of ethnology and arts and crafts. By the turn of the century the construction of native villages was a feature of ethnographic display at international exhibitions, and a Māori *pā* or fort at the Christchurch exhibition was likewise intended to be a 'picturesque' spectacle for visitors, a contrast between the stone-age Māori and the technological achievements of the young nation. However, the evidence of Māori involvement suggests that something quite different happened. From Māori politician James Carroll, who funded the project, to carver Hori Pukehika, who constructed it and feisty Ngāti

Porou chief Tuta Nihoniho, who 'took over' the running of the *pā*, the experience shows how Pākehā objectives were mediated by Māori.[31]

The Māori *pā* has been frequently cited in critical postcolonial accounts of the Pākehā construction of traditional Māori art, focusing on Hamilton's attempts to construct an archaic museum-based representation of Māori culture that denied innovation.[32] However, the indigenous artists did not always perform according to the script. When we consider the carvers' creative license, deviating from the prescribed models of 'traditional' art, and the performers' modern adaptations of music, performing arts and costume frowned upon by ethnologists, it is no wonder Hamilton privately confessed that the result was 'not as strictly correct as I should wish'.[33] Outside the somewhat restricted air of the official exhibition, Māori agency was even more in evidence in the commercial sphere. Henry Uru, a local Māori who was also on the exhibition committee, operated a stall at the exhibition grounds and an emporium downtown where he sold a range of arts and crafts. He evidently saw no problem with selling both traditional and modern objects such as feather cloaks, woven tea-cosies and carved pipes. 'Māori curios! Māori curios!' his catalogue advertisement shouted, exhorting tourists and visitors to the exhibition to 'call and see the unequalled display of … Māori curios'.[34] An even more striking example of indigenous entrepreneurship was the village with carved houses and performing troupe organized by Maggie Papakura, or Makereti, a famous bilingual tour guide who, perhaps inspired by her involvement in the Christchurch exhibition, travelled to Australia and England in 1910–11 presenting Māori music and crafts to an international paying audience.[35] Native agency of this kind was of course not confined to New Zealand. Thomas gives numerous other examples from the Pacific Islands, while in northeast North America Iroquois bead workers sold souvenir crafts to Europeans at the end of the nineteenth century. These workers, as Phillips shows, were not just poor hawkers but women who were continuing a pre-contact practice of trade and travel that sustained practices, maintained identities and resisted colonial administration.[36]

After the First World War, as the Arts and Crafts movement advanced from the fringes into the centre of New Zealand artistic culture, an environment was created that made possible the elevation of Māori arts and crafts, a syncretic indigenous/adapted blend of material culture that in its formal qualities most closely resembled western types of art/craft production. There are many examples of this heightened awareness in the writings of artists, writers, scholars and critics in this period, ranging from ethnographic enthusiasm at one end to full-blown fine-art appreciation at the other. Scholar Elsdon Best at the Dominion Museum, the Māori-speaking Pakeha who wrote numerous volumes of ethnographic studies on 'the Māori as he was', regarded Māori objects largely as advanced examples of neolithic technology. However, he wrote that the Māori was 'more than an artisan, he was an artist … [who] combined the useful and the ornamental wherever it was possible'.[37] One prominent Māori enthusiast was Dr Peter Buck (known as Te Rangihiroa within his Taranaki tribe), who later became director of the Bishop Museum in Honolulu and Visiting Professor of Anthropology at Yale. Buck was more prepared to use the term 'decorative art' in his articles on latticework and weaving that appeared in scholarly journals.[38] The new generation of professional anthropologists, such as Cambridge-trained H.D. Skinner at Otago Museum, were more interested in the expressive role of cultural objects. In the *Museums Journal* in 1917, Skinner called carving 'savage art'.[39] In a public lecture in 1919 he said that although Māori were ignorant of metal and pottery, 'their work has attained a high excellence'.[40] Raymond Firth, the New Zealand-born anthropologist who had trained in London with Malinowski, preferred to study Māori 'arts and crafts' in their social context:

> In order to appreciate the full value of the art, it is necessary to study it not only in museum show-cases where it is as a thing dead and set apart, but also as far as possible in its original and natural setting in the villages and homes of the people, where it is full of life and character.[41]

The School of Māori Arts and Crafts

The liveliness of native arts and crafts in its original setting, so admired by Pakeha experts such as Firth and Buck, can indeed be observed in the Arts and Crafts movement that flourished in Māori communities alongside its 'cross-over' appeal to the mainstream arts world. This movement was as much about politics as poetics. In the 1920s Māori crafts were wielded as a political tool in the campaigns of lawyer and MP Āpirana Ngata of Ngāti Porou, who worked within state institutions to further an agenda of social and cultural development. The Board of Māori Arts, one of Ngata's mechanisms for funding Māori projects, was required by legislation to 'foster and encourage the study and practice of Māori arts and crafts'.[42] For Māori, the arts, language and heritage were all linked with the quest for political autonomy in the face of colonization. It is important to state that craft's political agenda did not make it inauthentic or invalidate attempts at cultural invention; nor can Ngata and his generation be seen as 'collaborators' or the victims of manipulative Europeans. Instead, Māori arts and crafts must be understood as part of a reconstructed identity of indigenous nationhood forged within the colonial nation that Ngata called 'Māoritanga'.[43] There are parallels to this native artistic 'nationalism', in which material culture created and re-created indigenous identities.[44] Indigenous accommodations of and responses to western nation building and nationalism are also seen elsewhere, but the demographics of the New Zealand case, with a relatively large native population in a small country, mean that Māori efficacy was more apparent both inside and outside the colonial state.

This assertiveness can be observed in the project to build and decorate a special room within parliament buildings for the committee, which included the Māori MPs who oversaw the work of the Department of Native Affairs. The scheme was steered by Ngata, who drew on the expertise of artist James MacDonald from the Dominion Museum as well as tribal experts. The completed room was opened in 1922, and appears in a photograph (Figure 3.3) with sitting MPs, including the men who represented the four Māori seats, Ngata among them. It is a clear expression in three dimensions of what Ngata referred to as the adaptation of Māori traits to European civilization. The walls are defined by *poupou*, or carved slabs, separated by *tukutuku*, or latticework, as found in the interior of tribal meeting houses, but here the carving extends to door jambs, skirting board and window surrounds – essentially a standard European shell that has been indigenized by Māori decoration. The white plastered roof includes Māori motifs

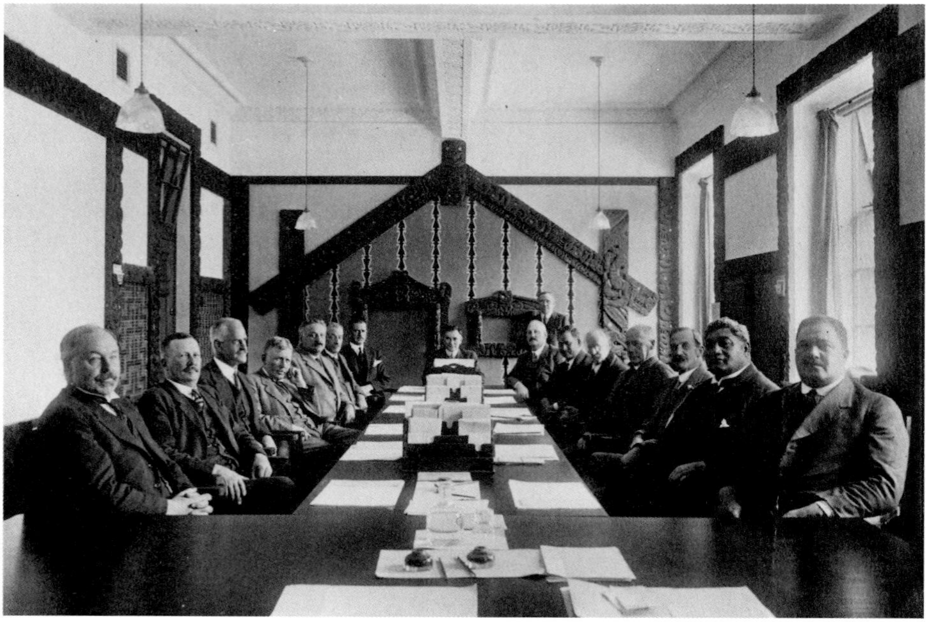

3.3 Photograph of the Native Affairs Committee in the Māori Affairs Committee Room, Parliament, Wellington 1922.

Note that Māori MP, Sir Āpirana Ngata, can be seen at the back centre second from the right.

while the pattern on the *tukutuku* is called *poutama* or 'the stairway to heaven', but it also has a modern interpretation to do with the acquisition of knowledge, and in this case may also have alluded to Māori social advancement. The most striking feature of the room is the outline of a *whare* or meeting house against the end wall, with the lines of the façade, door and window picked out in timber against the plain background. Here is a statement of the Māori relationship to the state, a house within *the* House, which may symbolize integration or separate development, depending on one's perspective. Ngata's own views may be divined from his later comments on the Māori house erected within the 1940 Centennial Exhibition building. He wrote that the 'blend of a beautiful ancient form with modern adaptations' reflects the 'Māori renaissance', and showed that 'the Māori people still retain after one hundred years an individuality of its own'.[45]

Ngata's major political achievement in the 1920s and 1930s was a series of social programmes in employment, education and land development, accompanied by cultural programmes. The flagship was the School of Māori Arts and Crafts established by an Act of Parliament in 1926. The School was backed by the Māori MPs, a sympathetic Prime Minister Gordon Coates, the Board of Māori Ethnological Research (set up by Ngata), the Dominion Museum and a tribal council, the Te Arawa Trust Board. Ngata's supporter and friend Gordon Coates, Minister of Native Affairs from 1921 to 1928, was one of those sympathetic Pākehā whom Ngata referred to as 'ngākau Māori' (Māori at heart).[46] A lively debate occurred in the House at the passing of the Bill, in which Pakeha politicians surprisingly expressed considerable support for the venture and saw Māori contributing to a national arts heritage. Though arguably an exercise in collaboration, the Māori input should not be underestimated and was motivated by deep-seated concern over the loss of skills in traditional crafts, a concern that had already prompted independent tribal initiatives.[47]

The initial proposal for the School of Māori Arts and Crafts argued that the best patron for the School was 'the Māori race'. Furthermore, at the first meeting of the Board, Te Arawa representatives insisted that its projects should be tribally based.[48] But those Māori involved were careful not to emphasize its tribal objectives and instead stressed the national mandate, the benefit for the whole nation. For example, the chairman of the Board of Māori Arts, MP and medical doctor Māui Pōmare, positioned their work within the mainstream ideology of nationhood. Phrasing his words selectively for a general audience, he was quoted in a newspaper as saying 'I think we can fairly claim to have the public with us in preserving the art of our Māoris'. 'Every country should try to preserve the original arts of its pioneers', he continued. It is noteworthy that Pōmare not only referred to 'our' Māori in the usual patronizing way that officials did, but employed the word 'pioneer' for indigenous people, not the Pākehā founding fathers of the settler nation – Māori were thus inscribed within the discourse of government as citizens and therefore deserving of state resources directed to preserving their culture.[49]

By the 1920s the category of arts and crafts became synonymous among Pākehā New Zealanders with the 'traditional' decorative arts of the *marae* (ceremonial meeting place). This shift can be detected in the positive press reaction to the setting up of the School of Māori Arts and Crafts in 1926–27, when phrases such as 'Māori crafts' and/or 'Māori art' often appear among similar terms, such as 'Māori artcraft', or 'native art'.[50] The phrase 'Māori arts and crafts' indicated a more general appreciation of cultural practices that were moving towards a pseudo-art status. For example, artist and technician James McDonald at the Dominion Museum is recorded as talking about 'the art of the kai-whakairo (carver)', 'decorative art', and 'Māori art and manufacture'. Yet the word 'art' in these constructions always seems to be qualified, and McDonald clearly preserved the distinction between European 'fine art' and Māori craft, and did not mix those objects in exhibitions at the museum.

However, a wide cross-section of Pākehā intellectuals had developed an appreciation for Māori arts and crafts. In 1928 the English artist and teacher W. Page-Rowe, working in Wellington and closely involved with the establishment of the School of Arts and Crafts, analysed carving in aesthetic terms – although he specified that it was applied rather than pictorial art. 'It is with Māori art as art', he said, 'with the artistry of Māori handiwork, that I am concerned'. Analysing a *tauihu* (canoe prow), he concludes that it is 'something very much more than an interesting ethnological specimen. It is a work of art'.[51] Many Māori seemed to have been flattered by this attention and readily agreed that Māori art was art. Others can be seen to pick up on particular strands of this sympathetic literature and, strategically at least, agreed with the public 'recognition' of the artistic nature of their material culture. Ngata's introduction to Page-Rowe's book, published by the Board of Māori Ethnological Research on which he sat, heralded the first serious study by an artist of 'Māori Art' (with an upper-case A).

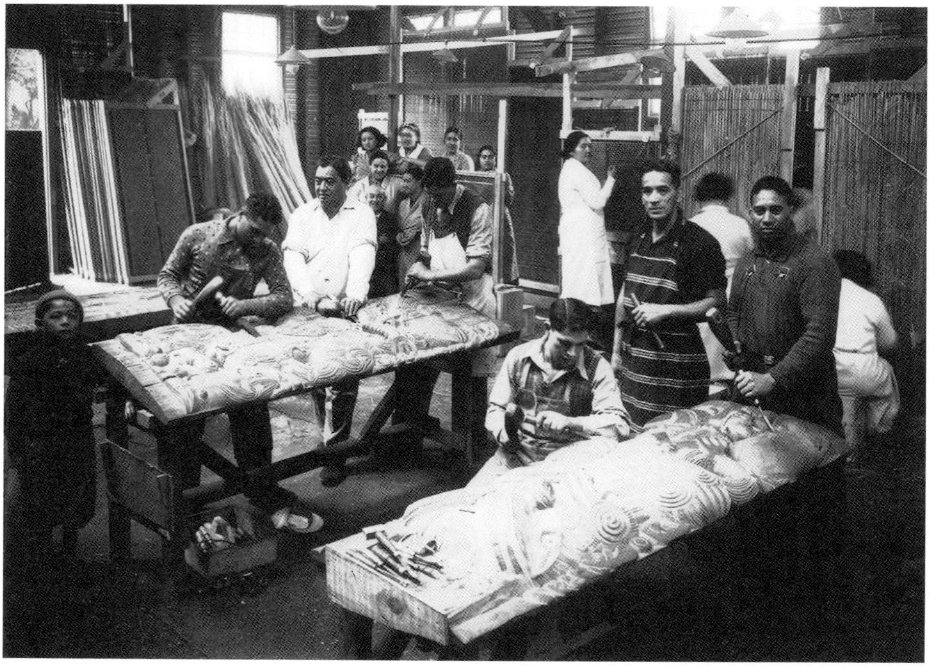

3.4 The carvers from the School of Māori Arts and Crafts at work at the Dominion Museum, Wellington, 1936.

In the background the museum carver Jacob Heberly can be seen with the women from the Ngāti Raukawa tribe working on *tukutuku* (lattice) panels.

Formerly, he noted, interest in 'Māori art objects' was ethnographic or commercial, but now with the new School the aim was 'the renaissance of the national art of the Māori, and not merely the craftsmanship'.[52] Although he was later criticized for imposing a traditional orthodoxy on carvers, initial ideas for the School were surprisingly open to contemporary influences and it was even envisaged that, although it would follow 'old time models and traditions' preserved in museums, Māori art would eventually develop in new directions.[53] Page-Rowe's proposal stated that 'the artistic genius of the Māori still survives' and that new applications and materials would not produce 'un-Māori works of art'.[54] Similarly James McDonald's proposals argued that 'out of present day influences, and the knowledge of the ancient arts and crafts, new forms, distinctive in their originality, would be created'.[55]

The operation of the School of Arts and Crafts has been described in detail elsewhere.[56] It was based in the thermal resort of Rotorua, where carving skills were maintained having been fostered by the souvenir trade for tourists. The School also moved around the country to regions where meeting houses, dining halls and memorials were being built in tribal communities. Young men were trained in chisel and adze work by the few experienced carvers remaining who were brought in as teachers. Figure 3.4 shows the young carvers from the School at work in the Dominion Museum in Wellington. In the foreground we see them working on two *poupou* or wall slabs of native *totara* timber destined for a house at Waitara and a memorial hall for Wairoa, while in the background a group of local women from Ngāti Raukawa tribe are completing *tukutuku* panels. Among them we see the seated figure of Thomas Heberly, the Māori carver who worked at the museum full time. While in the museum the carvers from the School studied the carvings in the collection closely as models for new carved houses in a 'traditional' style, while simultaneously 'restoring' the renowned meeting house *Te Hau ki Tūranga* as a centrepiece for the new museum building. Ngata's own research into the origins of carving concluded that the surviving portions of this famed house, created in the 1840s by the Rongowhakaata people from Gisborne, were 'the best extant examples of house carvings'.[57] As the museum was directly behind parliament buildings, Ngata was able to keep a close eye on the work, and in the evenings he would lead the group in *waiata a ringa* (action songs), which they would perform in public at concert halls to raise money, and on *marae* (community complexes) at the opening of their completed *wharenui* (big houses).

No fewer than 30 new '*whare rūnanga*' or meeting houses were built in the two decades the School was running. Although Howard Hamilton was ostensibly the head, Ngata was its driving force, coordinating projects, securing funding and constantly monitoring the standard of work. The underlying objective was to 'revive elements of Māori culture' associated with *marae*. In a letter to a friend, Ngata summarized the achievements resulting from their first carving project, the meeting house *Mahinārangi* at Ngāruawāhia in 1928–29. In the process of completing the first fully carved *whare* in the Waikato region in many years, local people were taught the arts of wood carving, rafter painting and plaiting wall panels; there was a new interest in plaiting *whāriki* (mats) and preparing the natural fibres *kiekie*, *harakeke* and *kākaho*; as well as a revival of traditional songs and incantations performed in opening ceremonies. The new house was an 'abiding inspiration'[58] to the Tainui tribes, and subsequently became the headquarters of the resurgent Māori King movement that still flourishes today.

The revival of crafts was clearly important to Māori people in terms of their own value systems. In a manuscript written in the 1930s, Pine Taiapa, the leading carver in the School, described how arts and crafts added to the *mana* of tribes: a complex concept meaning reputation, prestige, power and authority.[59] Another graphic example was a booklet produced by Native School teachers after a workshop with Taiapa to encourage carving among Māori children on the East Coast.[60] Each opening of a new or refurbished house was accompanied by extensive ceremony, performing-arts competitions, sports tournaments and other activities, all carefully orchestrated to revive traditional practices and engender pride in tribal identity. Ngata wrote that the spirit of 'Māoritanga', or Māoriness, is 'manifested in persistent culture traits, carvings, plaiting, marae ceremonial, poetry and song, speech and so forth'.[61] For example, after the opening of *Raukawa* at Ōtaki in 1936, Ngata explains the social value of this '*taonga*' to Ngāti Raukawa:

> They now possess something their hearts can embrace, and something which will help them to recall the spirit of their forebears, and a common meeting place for the hopes and aspirations of the younger generation of this Pakeha world.[62]

In a political pamphlet distributed to Ngāti Porou before the 1935 election, the Member for Eastern Māori told his constituents that he had worn himself out 'reviving the true works of our ancestors as *taonga* (treasures) for this generation'. Those *taonga* – language and literature (*kōrero*), performing arts (*haka* and *waiata*), carving, house building and the maintenance of *marae* customs, and the retention of *mana* (power, authority) and political independence (*rangatiratanga*) – represent an inventory of Māori cultural aspirations that continue to the present.[63]

'The finest specimens of Māori art and handicraft': Museum Display 1920s–1930s

If most Pākehā were unaware of these developments in Māori communities, many probably saw the Māori arts and crafts displayed in public museums and exhibitions in the 1920s and 1930s. A whole issue of the journal *Art in New Zealand* was devoted to 'Māori art' in 1929, containing articles by journalist James Cowan and Howard Hamilton, director of the School of Arts and Crafts. 'There was a time', remembered Cowan, when Māori art was associated with 'the grotesque and the barbaric'. Now the 'art culture' of the Māori was attracting the 'sympathetic study' of artists, architects and designers, who saw in the '*whakairo* of old New Zealand' a 'rich store' for 'national art ideals'.[64]

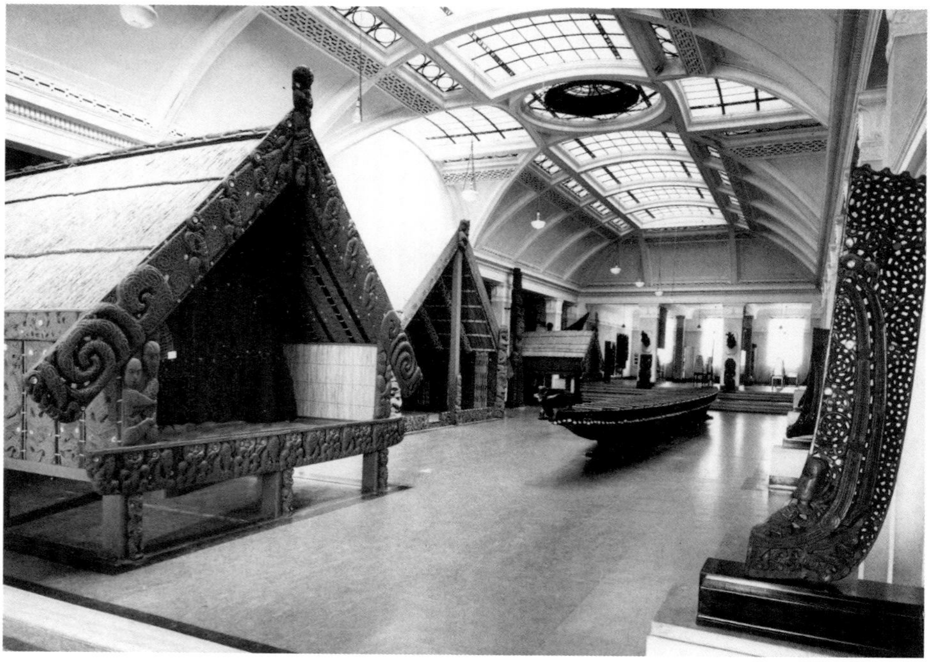

3.5 The interior of the Auckland War Memorial Museum built in 1929, showing the Māori court at the centre of the building with major carved artifacts: a *pātaka* or storehouse (left), *wharenui* or house (centre) and *waka* or canoe (right).

How was this rising appreciation reflected in public display? When the new Auckland War Memorial Museum opened in the country's largest city Auckland in 1929, Māori arts and crafts took pride of place in a central 'Māori court' (Figure 3.5). The European Arts and Crafts style and detailing of the architecture and decorative plasterwork, which incorporated Māori motifs, was matched by the richly detailed wood carving in the space itself where a number of large structures were brought together in a fanciful recreation of a Māori village: a *pātaka* or storehouse (left), *wharenui* or house (centre) and *waka* or canoe (right). Note how the curvilinear tracery and circular spirals of the *tauihu* (canoe prow) at the right are echoed by the curved shape of the roof and the central window in the skylight above. This new museum was opened with great fanfare, including a Māori ceremony led by the *kaumātua* (elders) of the Ngāti Maru people who sold the house *Hotunui*. Admittedly some of this acknowledgement of Māori culture may have been token and fanciful – the usual cutting of the ribbon was replaced by a dignitary knocking at the front door with a Māori *patu* (club) – yet the Museum *had* taken some steps to engage with Māori communities.

In the next decade the capital city of Wellington saw an even more grandiose temple erected to Māori arts and crafts. The Dominion Museum and National Art Gallery was opened in 1936, and together with the National War Memorial and Carillon, sited above the city on Mount Cook, was declared 'Wellington's Acropolis'. As visitors entered, the foyer opened out into the impressive space of the Māori Hall, with the magnificent meeting house *Te Hau ki Tūranga* at the centre facing them, flanked by *pātaka* and *waka* on either side, as in Auckland. The *Dominion* hailed it as a 'national icon of Māori artistry'.[65] The Governor General Lord Bledisloe praised the new Museum and its Māori displays as a 'perpetual reminder that this Dominion is the natural centre for the display of the finest specimens of Māori art and handicraft'.[66] The ethnologist on the staff, W.J. Phillipps, singled out *Te Hau ki Tūranga* as 'one of the most outstanding examples of Māori art workmanship in existence'.[67] By the 1940s, we can see clearly how Māori arts and crafts had risen in Pākehā estimation and become part of the nation's common heritage.

In fact archival research reveals that the restoration and installation of this renowned meeting house was the result of direct intervention by Ngata himself. Although staff plans for the exhibit were well advanced, one year from the opening in August 1935 Ngata told the director W.R.B. Oliver that he was 'desirous of having the Meeting House erected in true Māori style according to the customs of the East Cape district'.[68] In response, the Museum handed over responsibility to Ngata.[69] The museum carver Thomas Heberley and the team of carvers from the School of Arts and Crafts were based in the museum workshop and carving store in Sydney Street East to allow Ngata easy access from his office next door in Parliament buildings. The work proceeded well (Figure 3.4) and attracted so many curious visitors that it had to be closed to the public. Ngata's comments at the time suggest that the display of *Te Hau ki Tūranga* was intended to celebrate Māori arts and crafts in the new national museum and become a model for the newly traditional carved meeting houses springing up around the country.[70]

Conclusion

With Āpirana Ngata's guidance, the communal category of Māori arts and crafts entered the museum world through the analogous Arts and Crafts movement, like a Trojan horse in the citadel of European high culture. Despite the imperfect outcome (the interior of the house was not completed in time for the opening due to a shortage of materials), this was one of the first genuinely independent attempts by indigenous people to represent themselves in a museum. Arts and crafts were thus enshrined at the very heart of New Zealand museums just as, in the political and social field, Māori were trying to carve out a more equitable place for themselves within the Dominion of New Zealand, despite land loss, poverty and discrimination. As always, engaging with the settler state meant walking a fine line between assimilation and self-determination. But the Māori leaders who campaigned for their cultural heritage were determined and pragmatic people who wanted their way of life to survive and develop. At the same time as they wrestled with government structures, their material culture in the form of Māori arts and crafts found its way into the public domain, including museum display. Ngata and his generation clearly looked at museums, exhibitions, books and visual and performing arts as a means of uplifting Māori pride and declaring their presence on the national stage. There are vital lessons here for the analysis of indigenous material culture in settler colonies and the way it became entwined with European craft production. Māori negotiated with and through official policies and politicians to advance their own interests. The rising trajectory of Māori arts and crafts traced in this chapter illustrates how these practices can be seen not in terms of resignation, nostalgia for the past and cultural loss, but as symbols of survival, strength and confidence in the future.

Manuscript Sources Cited

Alexander Turnbull Library, Wellington, 'Nga taonga a Rauru-ki-tahi: Practical hints on Māori carving' (Whakawhitira Native School, Ruatoria, 1939), Pq731.462 PRA 1939.

Archives New Zealand, Wellington, Augustus Hamilton, 'Māori art: Should we teach it and apply it?' *Evening Post*, October 6, 1904, TO 1, 1901, 162/20.

Te Papa Archives, Wellington, Augustus Hamilton, letter to Elsdon Best, July 4, 1906, MU 152, 2/6.

Te Papa Archives, Wellington, Augustus Hamilton, letter to Minister Internal Affairs, May 29, 1911, Hamilton miscellaneous files, MU 206 1.

Te Papa Archives, Wellington, Āpirana Ngata and Herbert Williams, 'Proposal to establish a School of Māori Art', Board of Māori Ethnological Research, August 16, 1926, 'Māori Arts and Crafts Act', MU 14 7 (13/38/2).

Te Papa Archives, Wellington, 'Arts and crafts among the Māoris', *Evening Post*, September 21, 1928, 'Cuttings re museum and art gallery 1926–8', MU 7, 24/16.

Te Papa Archives, Wellington, James McDonald, 'Submitted for discussion at the next meeting of the Board of Māori Arts', April 5, 1927, MU 14 7.

Te Papa Archives, Wellington, Johannes Anderson, letter to Augustus Hamilton, June 17, 1906, Hamilton correspondence, MU 152, 1/3.

Te Papa Archives, Wellington, 'Māori art-craft: The proposed Rotorua school', *Auckland Star*, October 2, 1926, 'Māori clipping books 1910–51', MU 157, book 3.

Te Papa Archives, Wellington, 'School of native art', *Rotorua Chronicle*, January 7, 1927; 'Fostering Māori art'; *Herald*, April 6, 1927; 'Board of Māori Arts' MU 14, 7.

Hocken Collections, Otago University, Dunedin, H.D. Skinner, 'Notes from popular lecture on Anthropology Dunedin 1919', Ms 1219/083 'Ethnological, methods – historical'.

Printed Primary Sources Cited

Bledisloe, C.B. 'The Proper Function and Scope of a National Art Gallery and Museum', *Address given when laying the foundation-stone of the National Art Gallery and Dominion Museum at Wellington, April 14, 1934* (Auckland: Wilson & Horton, 1934).

Dominion, August 1, 1936.

New Zealand and South Seas Exhibition, Dunedin, 1889–90: Official Catalogue (Wellington: Government Printer, 1889).

New Zealand centennial exhibition: The Māori court. Souvenir (Wellington: Native Department, 1940), pp. 4–5, 11.

O'Rourke, Ross, 'Te Hau ki Turanga: A chronological document bank' (Wellington: Museum of New Zealand, 1994), Memo W.R.B. Oliver to Internal Affairs, August 14, 1935, n.p.

Sharpe, Alfred, letter to the editor, *The New Zealand Herald*, October 27, 1885, p. 6.

Secondary Sources Cited

Barringer, T.J., and Tom Flynn, *Colonialism and the Object: Empire, Material Culture, and the Museum* (London and New York: Routledge, 1998).

Belich, James, *Paradise Reforged: A History of the New Zealanders from the 1880s to the Year 2000* (Auckland: Allen Lane and Penguin, 2001).

Buck, Peter, 'The Evolution of Māori Clothing', *Journal of the Polynesian Society*, 33/129 (1924), pp. 25–47.

Buck, Peter, 'Māori Decorative Arts', *Transactions of the New Zealand Institute*, 53 (1921), pp. 452–70.

Calhoun, Ann, *The Arts and Crafts Movement in New Zealand, 1870–1940: Women Make Their Mark* (Auckland: Auckland University Press, 2000).

Cameron, Fiona, 'Shaping Māori Identities and Histories: Collecting and Exhibiting Māori Material Culture at the Auckland and Canterbury Museums from the 1850s to the 1920s', PhD thesis, Anthropology (Massey University, Palmerston North, 2000).

Condliffe, J.B., *Te Rangi Hiroa; the Life of Sir Peter Buck* (Christchurch: Whitcombe and Tombs, 1971).

Coombes, Annie, *Reinventing Africa: Museums, Material Culture and Popular Imagination in Late Victorian and Edwardian England* (London: Yale University Press, 1994).

Coombes, Annie (ed.), *Rethinking Settler Colonialism: History and Memory in Australia, Canada, Aotearoa New Zealand and South Africa* (Manchester: Manchester University Press, 2005).

Cowan, James, 'The art craftmanship of the Māori', *Art in New Zealand*, December, 2/6 (1929), pp. 121–31.

Diamond, Paul, *Makereti: Taking Māori to the World* (Auckland: Random House, 2007).

Firth, Raymond, 'The Māori carver', *Journal of the Polynesian Society*, 35/136 (1925), p. 83.

Gell, Alfred, *Art and Agency: An Anthropological Theory* (Oxford: Clarendon Press, 1998).

Hamilton, Augustus, *The Art Workmanship of the Māori Race in New Zealand* (Dunedin: New Zealand Institute, 1901).

Hamilton, George Heard, *The Pelican History of Art: Painting and Sculpture in Europe 1880–1940* (Harmondsworth: Penguin, 1978).

Henare, Amiria, *Museums, Anthropology and Imperial Exchange* (Cambridge: Cambridge University Press, 2005).

Jacknis, Ira, *The Storage Box of Tradition: Kwakiutl Art, Anthropologists, and Museums, 1881–1981* (Washington, DC: Smithsonian Institution Press, 2002).

Kaa, Wiremu, and Te Ohorere Kaa (eds), *Āpirana Turupa Ngata: Ana Tuhinga I Roto I Te Reo Māori* (Wellington: Te Whare Wananga o te Upoko o te Ika a Maui, Victoria University Press, 1996).

Kreps, Christina F., *Liberating Culture: Cross-Cultural Perspectives on Museums, Curation, and Heritage Preservation* (London and New York: Routledge, 2003).

Latour, Bruno, *Reassembling the Social: An Introduction to Actor-Network-Theory* (Oxford: Oxford University Press, 2005).

McCarthy, Conal, 'From curio to taonga: A genealogy of display at New Zealand's national museum, 1865–2001', PhD thesis, Museum and Heritage Studies (Victoria University, Wellington, 2004).

McCarthy, Conal, *Exhibiting Māori: A History of Colonial Cultures of Display* (Oxford and New York: Berg, 2007).

McCarthy, Conal, 'New Zealand and South Seas Exhibition, Dunedin, 1889–90', in J. Findling and K. Pelle (eds), *Encyclopedia of Worlds Fairs and Expositions* (Jefferson, NC: McFarland, 2008), pp. 108–10.

McCarthy, Conal, '"Our works of ancient times": History, colonisation and agency at the 1906–7 New Zealand International Exhibition', *Museum History Journal*, 2/2 (2009), pp. 119–42.

McCarthy, Conal, 'Carving out a place in the Better Britain of the South Pacific: Māori in New Zealand museums and exhibitions', in Sarah Longair and John McAleer (eds), *Curating Empire: Museums and the British Imperial Experience* (Manchester: Manchester University Press, 2012).

McNaughton, Howard, 'Mixing Metonyms: Staging Polynesia in the Christchurch International Exhibition 1906', in Howard McNaughton and John Newton (eds), *Figuring the Pacific: Aotearoa & Pacific Cultural Studies* (Christchurch: Canterbury University Press, 2005), pp. 83–97.

Mead, Sidney Moko (ed.), *Te Māori: Māori Art from New Zealand Collections* (New York: Heinemann & American Federation of Arts, 1984).

Neich, Roger, *Carved Histories: Rotorua Ngāti Tarawhai Woodcarving* (Auckland: Auckland University Press, 2001).

Ngata, Āpirana, 'Māori arts and crafts', in I.L.G. Sutherland (ed.), *The Māori People Today: A General Survey* (Wellington: New Zealand Council of Educational Research, 1940), pp. 307–35.

Ngata, Āpirana, 'The origin of Māori carving', *Te Ao Hou*, 6.2 (1958), pp. 30–37 and 'The origin of Māori carving', *Te Ao Hou*, 6.3 (1958), pp. 30–34.

Page-Rowe, W., *Māori Artistry* (New Plymouth: Board of Māori Ethnological Research, 1928).

Pishief, Elizabeth, 'Augustus Hamilton: Appropriation, ownership, authority', MA thesis, Museum Studies (Palmerston North: Massey University, 1998).

Phillips, Ruth B., *Trading Identities: The Souvenir in Native North American Art from the Northeast, 1700–1900* (Seattle, WA: University of Washington Press, 1998).

Phillips, Ruth B. 'Making Sense out/of the Visual: Aboriginal presentations and representations in nineteenth-century Canada', in D. Cherry (ed.), *Art: History: Visual: Culture* (Malden, MA and Oxford: Blackwell, 2005), pp. 113–35.

Phillipps, W.J., 'Māori carving', *The Public Service Journal*, March (1938), p. 167.

Sorrenson, M.P.K. (ed.), *Na to Hoa Aroha: From Your Dear Friend. The Correspondence between Sir Āpirana Ngata and Sir Peter Buck 1925–1950*, vol. 3, 1932–50 (Auckland: Auckland University Press, Māori Purposes Fund Board, 1988).

Skinner, Damian, *The Carver and the Artist: Māori Art in the Twentieth Century* (Auckland: Auckland University Press, 2008).

Taiapa, Pine, 'Education in a plural society', *Te Māori*, 2/6 (1972), pp. 35–7.

Thomas, Nicholas (ed.), *Rauru: Tene Waitere, Māori Carving, Colonial History* (Dunedin: Otago University Press, 2009).

Thomas, Nicholas, *Entangled Objects: Exchange, Material Culture, and Colonialism in the Pacific* (Cambridge, MA: Harvard University Press, 1991).

Thomas, Nicholas, *Possessions: Indigenous Art/Colonial Culture* (London: Thames & Hudson, 1999).

Thomson, G.E., *Official Record of the New Zealand and South Seas International Exhibition, Dunedin 1925–26* (Dunedin: New Zealand & South Seas Exhibition Company & Coulls, Somerville, Wilkie Ltd, 1927).

Thomson, John Mansfield (ed.), *Farewell Colonialism: The New Zealand International Exhibition Christchurch, 1906–7* (Palmerston North: Dunmore Press, 1998).

Walker, Ranginui, *He Tipua: The Life and Times of Sir Āpirana Ngata* (Auckland: Viking, 2001).

Walker, Ranginui, *Ka Whawhai Tonu Matou: Struggle without End* (Auckland: Penguin, 2004).

Willmott, Cory, 'Designing, producing and enacting nationalisms: Contemporary Amerindian fashions in Canada', in Beverly Lemire (ed.), *The Force of Fashion in Politics and Society: Global Perspectives from Early Modern to Contemporary Times* (Aldershot: Ashgate, 2010), pp. 167–90.

Notes

1 T.J. Barringer and Tom Flynn, *Colonialism and the Object: Empire, Material Culture, and the Museum* (London and New York, 1998).

2 Ruth B. Phillips, *Trading Identities: The Souvenir in Native North American Art from the Northeast, 1700–1900* (Seattle, 1998), p. 3.

3 James Belich, *Paradise Reforged: A History of the New Zealanders from the 1880s to the Year 2000* (Auckland, 2001), p. 190.

4 See, for example, Annie Coombes (ed.), *Rethinking Settler Colonialism: History and Memory in Australia, Canada, Aotearoa New Zealand and South Africa* (Manchester, 2005).

5 Ranginui Walker, *He Tipua: The Life and Times of Sir Āpirana Ngata* (Auckland, 2001) and *Ka Whawhai Tonu Matou: Struggle without End* (Auckland, 2004).

6 Ann Calhoun, *The Arts and Crafts Movement in New Zealand, 1870–1940: Women Make Their Mark* (Auckland, 2000).

7 John Thomson (ed.), *Farewell Colonialism: The New Zealand International Exhibition Christchurch, 1906–7* (Palmerston North, 1998). See also Calhoun, *The Arts and Crafts Movement in New Zealand, 1870–1940*, pp. 120–21.

8 Nicholas Thomas, *Possessions: Indigenous Art/Colonial Culture* (London, 1999).

9 Conal McCarthy, 'From Curio to Taonga: A Genealogy of Display at New Zealand's National Museum, 1865–2001', PhD thesis, Museum and Heritage Studies, Victoria University (Wellington, 2004).

10 Belich, *Paradise* Reforged, pp. 206–10. Conal McCarthy, 'Carving out a place in the Better Britain of the South Pacific: Māori in New Zealand museums and exhibitions', in Sarah Longair and John McAleer (eds), *Curating Empire: Museums and the British Imperial Experience* (Manchester, 2012).

11 Sydney Moko Mead (ed.), *Te Māori: Māori Art from New Zealand Collections* (New York, 1984), p. 21. See also Roger Neich, *Carved Histories: Rotorua Ngāti Tarawhai Woodcarving* (Auckland, 2001), pp. 123–37.

12 Neich, *Carved Histories*, p. 127.

13 Nicholas Thomas (ed.), *Rauru: Tene Waitere, Māori Carving, Colonial History* (Dunedin, 2009).

14 Damian Skinner, *The Carver and the Artist: Māori Art in the Twentieth Century* (Auckland, 2008).

15 Mead, *Te Māori*, pp. 22–4.

16 Amiria Henare, *Museums, Anthropology and Imperial Exchange* (Cambridge, 2005), p. 3.

17 Ibid., pp. 5–7. Alfred Gell, *Art and Agency: An Anthropological Theory* (Oxford, 1998). See also Bruno Latour, *Reassembling the Social: An Introduction to Actor-Network-Theory* (Oxford, 2005).

18 Conal McCarthy, *Exhibiting Māori: A History of Colonial Cultures of Display* (Oxford, 2007).

19 Alfred Sharpe, letter to the editor, *The New Zealand Herald*, October 27, 1885, p. 6.

20 Elizabeth Pishief, 'Augustus Hamilton: Appropriation, Ownership, Authority', MA thesis, Museum Studies, (Massey University, Palmerston North, 1998), p. 4.

21 *New Zealand and South Seas Exhibition, Dunedin, 1889–90: Official Catalogue* (Wellington, 1889) p. 177. See also Conal McCarthy 'New Zealand and South Seas Exhibition, Dunedin, 1889–90', in J. Findling and K. Pelle (eds), *Encyclopedia of Worlds Fairs and Expositions* (Jefferson, NC, 2008), pp. 108–10.

22 Augustus Hamilton, *The Art Workmanship of the Māori Race in New Zealand* (Dunedin, 1901). Calhoun, *The Arts and Crafts Movement in New Zealand, 1870–1940*, p. 89.

23 Pishief, 'Augustus Hamilton', pp. 35–8.

24 Annie Coombes, *Reinventing Africa: Museums, Material Culture and Popular Imagination in Late Victorian and Edwardian England* (London, 1994), pp. 143–4.

25 Johannes Anderson letter to Hamilton, June 17, 1906. Te Papa Archives, Wellington, Hamilton correspondence, MU 152, 1/3.

26 Augustus Hamilton letter to Minister Internal Affairs, May 29, 1911. Te Papa Archives Hamilton miscellaneous files, MU 206 1.

27 Augustus Hamilton, 'Māori art: Should we teach it and apply it?' *Evening Post*, October 6, 1904. Archives New Zealand TO 1, 1901, 162/20.

28 George Heard Hamilton, *The Pelican History of Art: Painting and Sculpture in Europe 1880–1940* (Harmondsworth, 1978), p. 131.

29 Hamilton, 'The Māori of New Zealand', undated typescript, Betty McFadgeon files, Te Papa Archives MU 47, 1/3.

30 McCarthy, *Exhibiting Māori*, pp. 92–9.

31 Conal McCarthy, '"Our Works of Ancient Times": History, Colonisation and Agency at the 1906–7 New Zealand International Exhibition', *Museum History Journal*, 2/2 (2009), pp. 119–42.

32 Howard McNaughton, 'Mixing Metonyms: Staging Polynesia in the Christchurch International Exhibition 1906', in Howard McNaughton and John Newton (eds), *Figuring the Pacific: Aotearoa & Pacific Cultural Studies* (Christchurch, 2005), pp. 83–97.

33 Augustus Hamilton letter to Elsdon Best, July 4, 1906. Te Papa Archives, Wellington, MU 152, 2/6.

34 Laura Fair, 'Colonial Curation through to Curation with Collaboration: A Study of the Institutional Culture within the Museum of New Zealand and Its Changing Approach to the Research, Collection and Display of Māori Culture from Its Formation through to the Present Day', MA thesis, History (Victoria University, Wellington, 2004), p. 85, figure 2.8.

35 Paul Diamond, *Makereti: Taking Māori to the World* (Auckland, 2007).

36 Nicholas Thomas, *Entangled Objects: Exchange, Material Culture, and Colonialism in the Pacific* (Cambridge, MA: Harvard University Press, 1991); Phillips, 'Making Sense out/of the Visual', pp. 256–7.

37 Elsdon Best quoted in G.E. Thomson, *Official Record of the New Zealand and South Seas International Exhibition, Dunedin 1925–26* (Dunedin, 1927), p. 83.

38 Peter Buck, 'The Evolution of Māori Clothing', *Journal of the Polynesian Society*, 33/129 (1924), pp. 25–47; 'Māori Decorative Arts', *Transactions of the New Zealand Institute*, 53 (1921), pp. 452–70.

39 Fiona Cameron, 'Shaping Māori Identities and Histories: Collecting and Exhibiting Māori Material Culture at the Auckland and Canterbury Museums from the 1850s to the 1920s', PhD thesis, Anthropology (Massey University, 2000), p. 185.

40 H.D. Skinner, 'Notes from popular lecture on Anthropology Dunedin 1919', Hocken Collections, Otago University, Dunedin, Ms 1219/083 'Ethnological, methods – historical'.

41 Raymond Firth, 'The Māori carver', *Journal of the Polynesian Society*, 35/136 (1925), p. 83.

42 *The New Zealand Times*, September 6, 1926.

43 Āpirana Ngata, 'Māori arts and crafts', in I.L.G. Sutherland (ed.), *The Māori People Today: A General Survey* (Wellington, 1940), p. 322.

44 On Australia and New Zealand, see Thomas, *Possessions*. For Canada, see Ira Jacknis, *The Storage Box of Tradition: Kwakiutl Art, Anthropologists, and Museums, 1881–1981* (Washington, DC, 2002). For an international overview see Christina Kreps, *Liberating Culture: Cross-Cultural Perspectives on Museums, Curation, and Heritage Preservation* (London and New York, 2003). For a contemporary example see Cory Willmott, 'Designing, producing and enacting nationalisms: Contemporary Amerindian fashions in Canada', in Beverly Lemire (ed.), *The Force of Fashion in Politics and Society: Global Perspectives from Early Modern to Contemporary Times* (Aldershot, 2010), pp. 167–90.

45 *New Zealand centennial exhibition: The Māori court.* Souvenir (Wellington: Native Department, 1940), pp. 4–5, 11.

46 J.B. Condliffe, *Te Rangi Hiroa; the Life of Sir Peter Buck* (Christchurch, 1971), p. 150.

47 Āpirana Ngata letter to Peter Buck, August 17, 1935, in M.P.K. Sorrenson (ed.), *Na to Hoa Aroha: From Your Dear Friend. The Correspondence between Sir Āpirana Ngata and Sir Peter Buck 1925–1950 Vol. 3, 1932–50* (Auckland: Māori Purposes Fund Board, 1988), p. 193.

48 Āpirana Ngata and Herbert Williams, 'Proposal to establish a School of Māori Art', Board of Māori Ethnological Research, August 16, 1926. TPA 'Māori Arts and Crafts Act', MU 14 7 (13/38/2). For Te Arawa involvement, see Ngata, 'Māori arts and crafts', p. 323.

49 'Arts and crafts among the Māoris', *Evening Post*, September 21, 1928, Te Papa Archives, 'Cuttings re museum and art gallery 1926–8', MU 7, 24/16.

50 'Māori art-craft: The proposed Rotorua school', *Auckland Star*, October 2, 1926; Te Papa Archives, 'Māori clipping books 1910–51', MU 157, book 3; 'School of native art', *Rotorua Chronicle*, January 7, 1927; 'Fostering Māori art', *Herald*, April 6, 1927; Te Papa Archives, 'Board of Māori Arts', MU 14, 7.

51 W. Page-Rowe, *Māori Artistry* (New Plymouth, 1928), pp. xi, 37.

52 The phrase was a quotation from Page-Rowe. See ibid., p. v.

53 Ngata and Williams, 'Proposal', p. 4.

54 Page-Rowe, 'Proposed Māori school of art', February 2, 1927. TPA MU 14 7.

55 James McDonald, 'Submitted for discussion at the next meeting of the Board of Māori Arts', April 5, 1927, Te Papa Archives MU 14 7.

56 Neich, *Carved Histories*; Skinner, *The Carver and the Artist.*

57 Āpirana Ngata, 'The origin of Māori carving', *Te Ao Hou*, 6/2 (1958): pp. 30–37 and 'The origin of Māori carving', *Te Ao Hou*, 6/3 (1958): pp. 30–34.

58 Āpirana Ngata letter to Peter Buck, August 17, 1935, in Sorrenson, *Na to Hoa Aroha*, p. 193.

59 Pine Taiapa, 'Education in a plural society', *Te Māori*, (1972): pp. 35–7.

60 'Nga taonga a Rauru-ki-tahi: Practical hints on Māori carving' (Whakawhitira Native School, Ruatoria, 1939), Alexander Turnbul Library, Wellington, Pq731.462 PRA 1939.

61 Ngata letter to Buck, June 11, 1933, in Sorrenson, *Na to Hoa Aroha*, p. 85.

62 Ngata letter to Buck, March 28, 1936, in ibid., p. 215.

63 Āpirana Ngata, 'Panui mo te pooti o te tau 1935', in Wiremu and Te Ohorere Kaa (eds), *Āpirana Turupa Ngata: Ana Tuhinga i roto i te Reo Māori* (Wellington, 1996), p. 393.

64 James Cowan, 'The art craftmanship of the Māori', *Art in New Zealand*, December, 2/6 (1929), p. 121.

65 *Dominion*, August 1, 1936.

66 C.B. Bledisloe, 'The Proper Function and Scope of a National Art Gallery and Museum', Address given when laying the foundation-stone of the National Art Gallery and Dominion Museum at Wellington, April 14, 1934 (Auckland: Wilson & Horton, 1934), p. 12.

67 W.J. Phillipps, 'Māori carving', *The Public Service Journal*, March (1938), p. 167.

68 Memo W.R.B. Oliver to Internal Affairs, August 14, 1935, in Ross O'Rourke, 'Te Hau ki Tūranga' (Wellington, 1994), n.p.

69 Memo W.R.B. Oliver to Internal Affairs, April 24, 1936, in ibid., n.p.

70 'Te Haukiteranga: Famous Māori house for new museum', *Dominion*, May 5, 1936.

4

Reconstructing the Old Dutch House: New Castle's Colonial Dutch Gem and Louise Crowninshield's Preservation Community

Alena Buis

> With its eaves only a few feet above the sidewalk, this quaint little house was built about 1665. Few facts are known of its history, but the lack of this knowledge leaves it none the less interesting, and perhaps even adds to its charm and old-world attraction.[1]

> It is the power of select objects to extend humans into larger communities of historical belonging.[2]

In a lecture on the challenges facing historical preservationists, Louise du Pont Crowninshield explained:

> The public won't go see a house just because of its architectural features … It must have a certain charm – the rouge pots and sewing baskets and all the little things through which today's people feel they are connected to yesterday's people … Attention to detail, therefore, is a constant necessity.[3]

Crowninshield was part of a community of Colonial Revivalists active in furnishing the Dutch House, a historic dwelling now touted by the New Castle Historical Society as Delaware's 'Early Colonial Gem' (Figure 4.1).

This essay considers the process of furnishing this house museum as a complex act of community building. I investigate early twentieth-century idealizations of early modern colonial communities and also the gendered communities formed through Colonial Revival collecting by focusing on the gathering of early colonial material culture to tell a larger story of Delaware's historic Dutch House originally constructed c. 1690–1710.

Today, the Dutch House is very much an example of one of Diane Barthel's *staged symbolic communities*: 'All communities manipulate symbols to define their boundaries

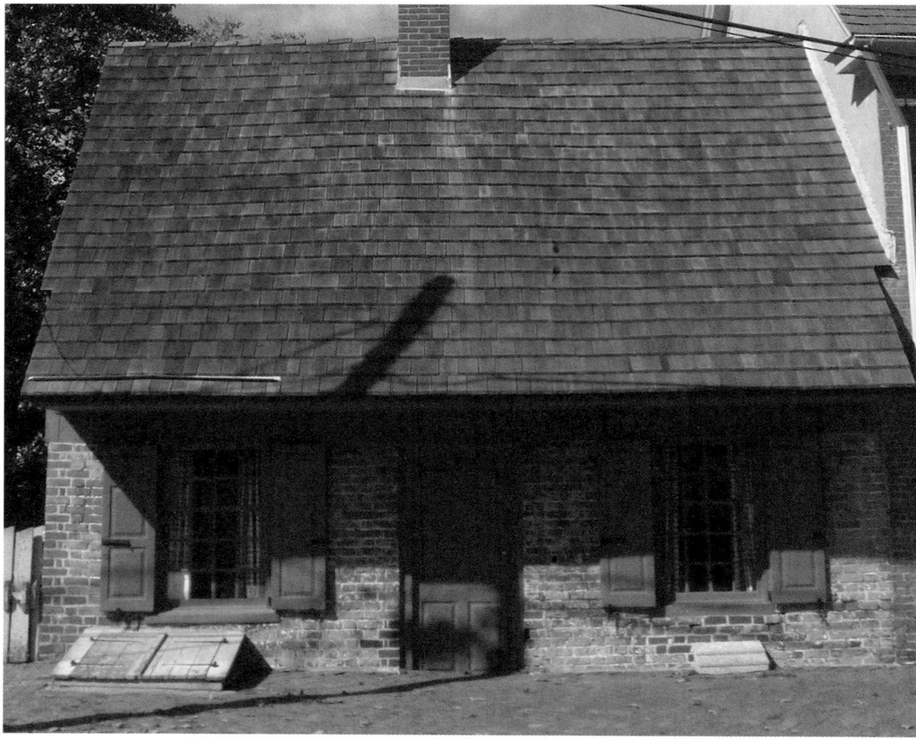

4.1 The Dutch House, New Castle, Delaware, 21 September 2012.

and identity'. In the case of settings like the Dutch House, 'these preserved or re-created villages are *themselves* symbols: they perform community in a society where organic communities are a thing of the past, if indeed, they ever existed'.[4] In 1937 the Dutch House was purchased by the Delaware Society for the Preservation of Antiquities (DSPA) and was later opened as a historic house museum. Shortly after, the DSPA charged Louise du Pont Crowninshield with assembling and organizing the display of objects. Led by Crowninshield, but involving many people with ties to the Colonial Revival movement, the social construction of the Dutch House is an example of the staged and imagined symbolic community identities created at the time. Furnished with European and American decorative art, the small house is considered to contain two intertwined narratives: primarily the town's early Dutch history, but secondarily it considers the ways collectors and preservationists of the early twentieth century interpreted early colonial life.[5]

Delaware's Early Colonial Gem

> [The] Old Dutch House, 3rd Street, between Harmony and Delaware Streets, facing the Green, is a small brick dwelling with low pent eaves projecting so far over the front of the house as to give the whole a storybook air, associated with tales and personalities, curious and fantastic. No one knows what span of life the little old house has seen.[6]

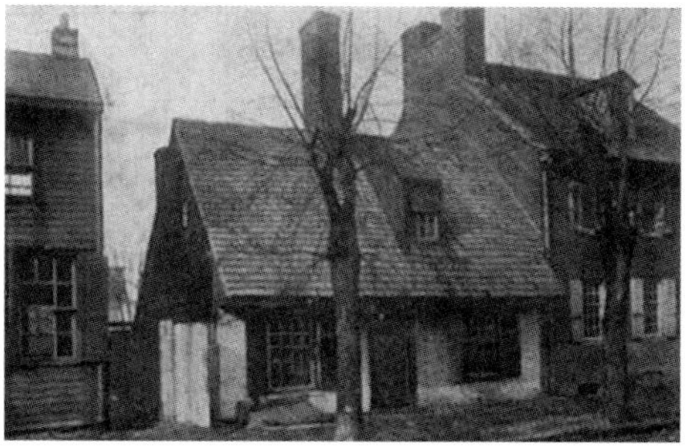

4.2 The Dutch House, New Castle, Delaware, undated postcard.

Throughout the nineteenth century the little building was referred to as the 'Old Dutch House', hinting at its colonial past. The original descriptors 'Old' and 'Dutch' were almost redundant because by the end of the nineteenth and beginning of the twentieth century, the adjective 'Dutch' referred to long-past colonial time.[7] The insistence on colonial origins was even reinforced by a deed record that stated: 'All that parcel of land, with the ancient brick building thereon erected, believed to be of Swedish or Dutch erection, situate in the City ... of New Castle [sic]'.[8] However, no Dutch names appear in its title provenance and it does not closely resemble any other easily attributable New Netherlandish architecture. The title of 'Dutch' is instead derived from the fact that it appears to be different from contemporary English models.[9] One 1936 report made for the Historic American Buildings Survey by Laussat Rogers concluded its Dutch origins based on the steep pitch of the roof and central chimney despite the fact that these both were most likely nineteenth-century additions (Figure 4.2).

Although the moniker of the Dutch House has persisted, the earliest indications of a building on the property fall outside Dutch possession of the colony of New Netherland from 1624 to 1664 (and brief reclamation in 1674) and significantly outside of the original Dutch West India Company (WIC) settlement of Fort Casimir established in 1651. In 1654, Swedes from nearby Fort Christina captured the settlement and renamed it Fort Trinity, only to have it recaptured by the Dutch in the following year. For almost ten years the small garrison was called Fort Amstel and controlled by the Dutch before finally becoming New Castle under English control; it remained so until the founding of the United States of America in 1776. The Dutch House materially contains traces of this multicultural history.

With few written records dating to before 1751, architectural historians have tentatively attributed the original timber-framed, one-room, single-story house to builder John Walker before 1696 or to Powell Barnes shortly after 1701. As in many early colonial structures, the internal framing is wood, and some of this finished and exposed interior and exterior wood siding can be seen on the gable end. Later, c. 1720, additions included a lean-to running the length of the back wall and the division of the large halls into two rooms and a tight, wound staircase to a half-story garret. In the mid-eighteenth century substantial renovations were made as the house was raised, a cellar excavated and the brick added to the exterior. By 1823, the Dutch House appeared similar to other houses surrounding it on Third Street, with a reconstructed second floor, Federal-style fireplaces, woodwork and floor plan. Early twentieth-century renovations undertaken by DSPA returned the building to its 1823 state.

In the most recent historical structures report on the Dutch House, Jeff Klee notes that 'Despite a contemporary, Colonial Revival mania among restoration architects for stripping away all historical layers except the earliest – and occasionally inventing an earliest layer', the architect hired by the DSPA, Albert Kruse, 'worked with a light touch, only making changes where structural necessity demanded'. Klee further noted that with evenly spaced H-bent structural frames typical of Netherlandish construction as well as the materials, spacing and manner of supporting the walls and floors in a typically English way, the frame of the Dutch House is neither English nor Dutch – it does not fit neatly into any known European typology. As such, it is an excellent example of how traditional architectural practices were translated and adapted to the differing physical and cultural conditions of the New World. A truly hybrid architecture was produced when 'the range of European architectural and cultural traditions in early colonial New Castle enabled the sharing of knowledge and fostered a climate of innovation that could produce an anomalous, un-classifiable structure like the Dutch House'.[10] This complex reading of a pluralistic colonial past disrupts many of the well-ordered, culturally homogeneous narratives of American colonial history.

The Tile House, another New Castle historic building on the Strand, now destroyed, had the same conjectural past as the Dutch House. As early as the 1820s one observer, John Watson, had noted that the irons embedded in the front of the building that proclaimed the 1687 construction gave it an 'aura of timeworn nobility'.[11] In the various artistic imaginings of local artists, the irons figure prominently despite their potential inaccuracy. Robert Montgomery Bird's 1826 watercolours of the building show it to be similar to other 'Dutch' buildings constructed throughout the territory of New Netherland. With its steep pitched roof and stepped gables, the Tile House was a relic of a far less heterogeneous architectural movement.[12] When John Boyer purchased the property in 1687, the building did not exist but a large brick house showed on the plot plans when his daughter inherited it in 1704. Benjamin Latrobe's rendering of the Tile House in his 1804–5 elevation survey does not include the irons; most likely the numbers were added as part of a later nineteenth-century renovation. This inaccuracy did not seem to sway New Castle residents, who viewed the destruction of the Tile House as a great loss to the community. This later fuelled the preservation of other historic buildings including the Dutch House and led to the founding of the DSPA in early 1937 in an effort to preserve the remaining architecture of New Castle.

Inspired by John D. Rockefeller's efforts in Virginia, the Society's initial goal was to create a similar 'Williamsburg on the Delaware'.[13] Ultimately the DSPA's only undertaking was to preserve the Old Dutch House. In July 1938 Colonel Daniel Moore Bates, committee chairman of the Society and neighbour of Louise Crowninshield, wrote a letter to another member explaining that for a sum they could obtain the building. The committee and Delawareans raised over $10,000, to purchase the house for $2,500 and also to repair the building's foundation and make other structural restorations. The project would be the DSPA's only major contribution to the preservation of New Castle; the outbreak of the Second World War prevented further efforts. In 1949 DSPA gave the Dutch House and its other properties and assets to New Castle Historical Society (NCHS).[14] Now owned and operated by the NCHS, the Dutch House is currently furnished much as it was during Louise Crowninshield's involvement.

Louise E. du Pont Crowninshield (1877–1958)

As a passionate participant of the early twentieth-century Colonial Revival movement and one of the leading early American historic preservationists, Louise Evelina du Pont Crowninshield is now best known for her participation in restoring and decorating her family's original stead of Eleutherian Mills in Delaware. Her contribution to the furnishing of George Washington Birthplace National Monument in Virginia is said to have 'heralded an entirely new role for the association while fundamentally redefining the function of historic objects at the birthplace'[15] (Figure 4.3).

Born in 1877 to Henry Algernon du Pont and Pauline Foster du Pont,[16] Louise and her younger brother Henry Francis du Pont were heirs to the du Pont industrial fortune. The two were born and raised at Winterthur, the family estate outside Wilmington, Delaware. Before her 1900 marriage to Boston Brahmin, Francis Boardman Crowninshield, her activities were typical for a wealthy young socialite: she mingled in New York and Washington, DC with the cities' most exclusive families and devoted her time to suitable charitable causes.[17] Throughout her life she was generous with her money as well as her time, serving on the boards of numerous museums, horticulture societies and cultural organizations.[18]

By 1949 Crowninshield helped to found the National Trust for Historic Preservation. She served on the Trust's board and, by the early 1950s, became president of the National Council for Historic Sites and Buildings when the two organizations merged. Addressing the contributions of the Trust, she revealed her approach to historic material culture:

> Americans will be better for having around them some visible remains of their past. Books and sketches and drawings and photographs will suffice [but] more thought than ever is being given to the three-dimensional documents of history. In the old that is good, there is more than mere patriotic reminder of the past; there is hope for the future.[19]

Despite her forays into male-dominated domains of art institutions, Crowninshield cannot be easily described as a feminist icon. Her success was achieved through her submission and the successful performance of circumscribed gendered roles.[20] In the early 1950s, when nominated to the presidency of the recently merged preservation committees, and after much urging, her response was hesitant: 'I don't think it is suitable; I am old-fashioned. I think a committee position should be held by a man'.[21]

Like the crafts that she collected, Crowninshield's vernacular, collaborative efforts have gone unrecorded in masculine histories that have marginalized women's participation. In her discussion of elite female collectors and their construction of dolls' houses, Susan Broomhall insists that the accumulated objects and management of space serve as an ego-document, providing insight into lives otherwise undocumented: 'Given the rich investment of energy and identity they demonstrate', perhaps we can add dolls' houses and their surrounding documentation as another form of ego-document that 'women could create as a testament to their interests and sense of self for their own era and for future generations'.[22] Crowninshield's work with the Dutch House serves as ego-documentation that locates her within preservation

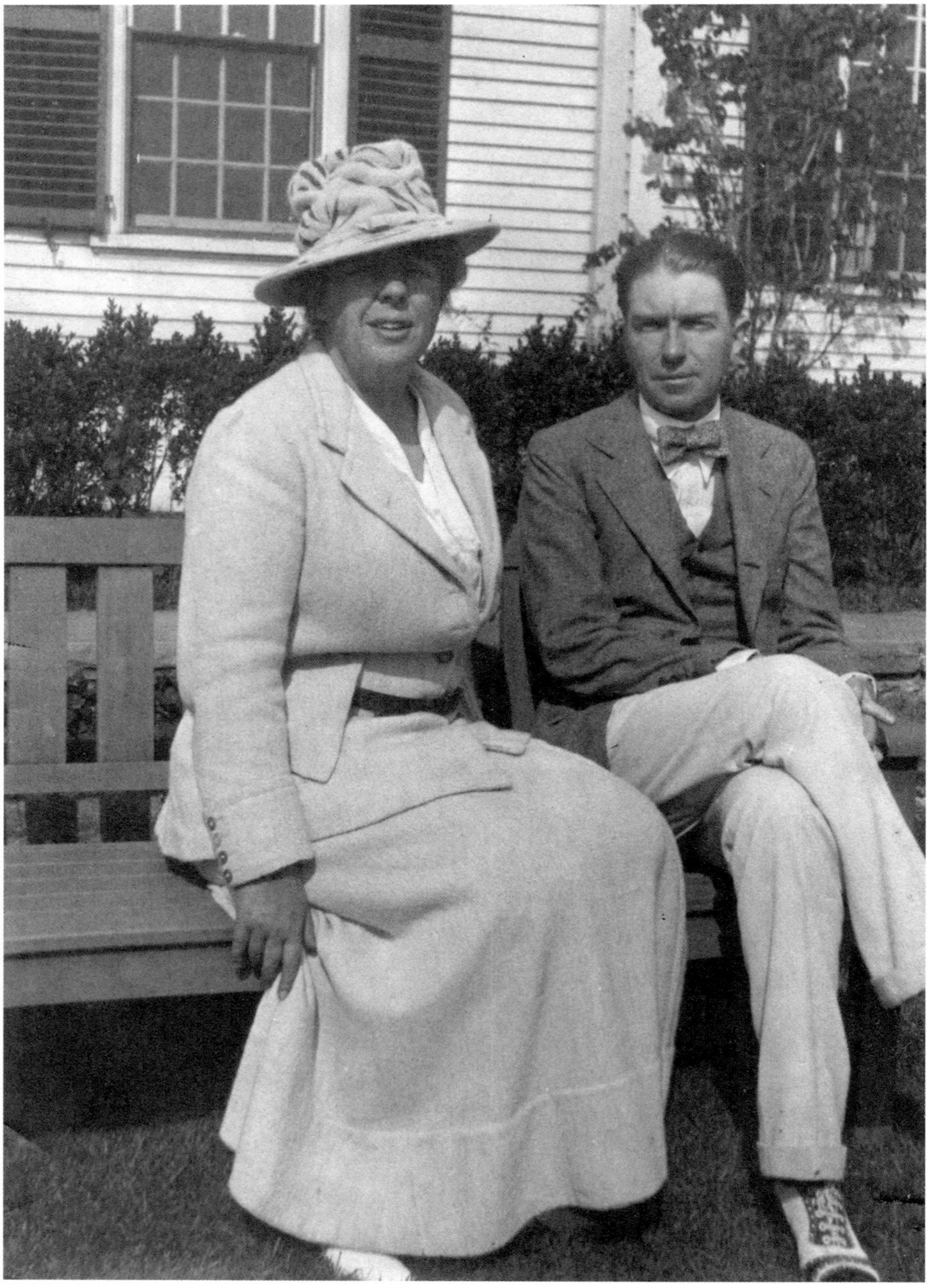

4.3 Photograph of H.F. du Pont and Louise Crowninshield.

communities. Not only is Crowninshield an unstudied figure in the history of historical preservation, but also the extent of her agency in projects such as the Dutch House articulates the gendered politics of early twentieth-century professionalization of cultural institutions.

Far less has been written about Crowninshield's collecting activities than about those of her brother Henry du Pont, although Crowninshield became interested in early American material culture first, and encouraged him to begin his own collections. A comparison of the du Pont siblings' collecting practices highlights how gender is 'expressed, shaped and marked through the process of collecting'.[23] Like her brother, Crowninshield was a member of the privileged elite who led a preservationist movement in which women of a particular class, like their male counterparts, had the social standing and economic means that allowed them access to vast cultural capital.[24] Their activities were further motivated by a sense of premier American citizenship characterized by their identification with their nationality, a sense of civic duty, interest in philanthropy and a desire to leave a legacy.[25] However, in contrast to her brother's well-documented activities,[26] primarily the founding of Winterthur Museum, Gardens and Library, home to one of the largest collections of American decorative arts, far less critical scholarship has focused on Crowninshield and her efforts. In *The Saturated World: Aesthetic Meaning, Intimate Objects, Women's Lives, 1890–1940*, Beverley Gordon briefly mentioned the contrasts between Crowninshield's activities for historic houses and her brother's accumulation at Winterthur as an example of the gendering of collecting practices and the ways in which Crowninshield has been marginalized in comparison to her more famed (male) sibling.

As is often the case with women's participation in cultural activities, it is not the lack of contribution but rather the ways in which women's histories are written that obscure their activities.[27] While men typically have been designated 'collectors', women have been discussed as 'consumers of objects', with their collections of textiles and ceramics subject to a hierarchy of the arts, in which paintings were deemed more aesthetically and historically valuable. Within gendered accounts of people's relationships to the material world women have been perceived as selecting objects for their decorative domestic potential or for sentimental reasons. As was the case with Crowninshield, a distinction was made between her methods and the systematic classification, scholarly interest or independent aesthetic value characteristic of (masculine) historical objects more commonly ascribed to du Pont's approach. Crowninshield was considered an 'antiquer' and her efforts were dismissed at the time as 'decorative' and therefore not worth recording.[28] Her participation in furnishing historic homes like the Dutch House were not characterized as serious, historically grounded endeavours, but rather viewed as an extension of women's prescribed domestic roles. This familial responsibility is emphasized in Crowninshield's participation in the restoration of Eleutherian Mills, the original du Pont family estate, where she was assigned the responsibility of a 'life tenant obligated to pass the house on intact, preserving all the valuable furnishings and works of art accumulated over decades or centuries'.[29] Before focusing on museums for the public, Crowninshield's initial projects centred on her own homes.

While du Pont's collection became monolithic, housed under a single roof that became a museum, Crowninshield's work was largely unacknowledged and, additionally, scattered throughout various locations with her participation credited only in the provenance. Whereas du Pont's collecting work was made public and well documented, Crowninshield's numerous projects were often collaborative, unattributed and primarily behind the scenes. Gordon further notes that du Pont's collection typified masculine accumulation as a more encyclopaedic representation of periods and styles devoid of place, in contrast to Crowninshield, who made choices with a certain space in mind. The distinctions continue: whereas du Pont's was a monument to his own taste, she created memorials for others. He incorporated other structures into his; she worked with the architecture she was given. And perhaps most importantly, while he was concerned primarily with the aesthetic visual effect, she wanted her interiors to be comfortable, to be lived in. His were independent art objects; hers were decorative and functional or utilitarian accompanied by a domestic narrative. Gordon describes Crowninshield's process as one of contextual collecting – objects were viewed as a part of an interior's whole.[30]

Communities: Imagined and Staged

One early twentieth-century New Castle resident tellingly proclaimed, 'This [house] is always called the Dutch House ... its story for the most part is lost in the passage of time. The Architecture is quaint and unusual'.[31] The 'quaint' and 'unusual' architecture of the Dutch House made it a particularly appropriate focus for early twentieth-century ideas of Dutch colonial participation. Annette Stott has described the increased collecting of Dutch painting, the re-visioning of American history to include Dutch political and cultural roots, and complex appropriation of Dutch signs and symbols as a Holland Mania that swept through the United States.[32] Its proponents believed Dutch culture to embody:

> a rich tradition of art; quality workmanship; ideal folk life; middle-class patronage of the arts; an emphasis on the home as a simple, comfortable, clean, yet picturesque refuge; and a stable lifestyle achieved through the use of uncomplicated technology.[33]

However, while unique, Dutch culture was not viewed as so ethnically distinct as to challenge the established white hegemony. Sharing a similar physical landscape, social custom, Protestant religion, democratic republicanism and interest in personal freedoms, the image of the Dutch was easily appropriated in American founding mythology.

One of the contributors to Holland Mania was Crowninshield's friend Douglas Campbell. During the years he danced with her at cotillions,[34] he also was writing extensive tomes like *The Puritan in Holland, England and America: An Introduction to American History* (1892). This volume traced the genesis of the American Declaration of Independence and the Federal Constitution, two of the most important American legal documents, to the Declaration of Independence and the Constitution of the United Provinces of the Netherlands. From the 1880s and well into the 1930s, the Netherlands became a popular resource in an American search for a national past and also for a contemporary cultural identity. The idealized Dutch founding mythology proposed by Holland Mania created what Benedict Anderson has termed an 'imagined community', whereby a nation and its colonial projects were socially constructed through the imagining of a particular group.[35] In the same vein, the Colonial Revival movement was 'an attitude or a mental process of remembering and maintaining the past that generations of Americans have quite consciously created'.[36]

Although in some ways the arrangements of furnishings for Colonial Revival period rooms and house museums have been interpreted to serve as 'a vehicle for the promotion of anti-modern, rather than progressive ideals',[37] the rhetoric behind the assimilation of recent immigrants was quite modern in its drive to create a homogenized, unified national culture. In his 1890 speech to the Holland Society, Theodore Roosevelt epitomized the contemporary understanding of seventeenth-century Dutch colonists as having:

> possessed the faculty of becoming good American citizens … pressing all their energies into the service of the mighty Republican Commonwealth … has a direct bearing upon one of the great questions of the present day, the assimilation of the masses of our foreign-born fellow citizens.[38]

The curatorial programming of house museums contributed to this ideology. For example, the Dyckman House, an early Dutch dwelling in Manhattan, was used as part of a 'kindergarten of citizenship' to provide young citizens with a 'living history of honest and upright life'.[39] From their viewing of seventeenth-century genre scenes, Americans stereotyped Dutch colonial homes as 'interiors that were filled with light and air, spotlessly clean, simple but richly furnished, and adorned with inspiring art'.[40] These characteristics had corresponding moral and philosophic implications: ideally American citizens were to be prosperous, clean and frugal. During a time of increased immigration to the United States, Crowninshield's involvement in preserving and furnishing the Dutch House was in line with Colonial Revival aims to assimilate a diverse ethnic population by creating a singular American cultural pride, modelled on a Dutch colonial precedent. With their charming material culture and quaint architecture, historic houses were viewed as valuable pedagogical tools in the assimilation of the population into one idealized community.

If businessmen of the Gilded Age found kindred spirits in the seventeenth-century burgher,[41] the Holland revival inundated their wives with idealized images of the ideal Dutch *huis vrouw*. Nowhere was the conflation of women and the interiors of homes more emphasized than in turn-of-the-century nostalgic visions of Dutch and Dutch-colonial domesticity.[42] Literature by Esther Singleton and Alice Morse Earle that combined studies of New York inventories with Dutch genre paintings emphasized a close association between Dutch material culture and that of colonial New York in a way that positioned the collecting of early American furniture as fulfilling a patriotic agenda.[43] And by focusing on hearth and home, women were able to establish the importance of maternal and domestic roles in foundational national mythologies. As a socially acceptable platform for women's agendas, historic house museums and period rooms were a way of making public what their curators had long expressed privately.[44] By 'inventing traditions' Crowninshield and other Colonial Revival communities employed:

> a set of practices, normally governed by overtly or tacitly accepted rules and of a ritual or symbolic nature, which seek to inculcate certain values and norms of behaviour by repetition, which automatically implies continuity with the past.[45]

Furnishing the Dutch House

When the DSPA approached her to assist with the furnishing of the Dutch House, Crowninshield was 'a rising star in the world of historic preservation', excelling at 'furnishing buildings with objects that, although not necessarily original to the building, were nonetheless perfect substitutes'.[46] The 1938 minutes of the DSPA note:

> Mrs Crowninshield reported that she was gathering furniture for the Old Dutch House. She said that she was strenuously opposed to placing a single piece of furniture in the house, unless it was the proper type and period. On motion of Mr Bissell it was voted to give Mrs Crowninshield full powers to furnish the house at her own discretion.[47]

But what was the 'type' and 'period' that Crowninshield had in mind? Due to the 'Holland Mania' influencing early twentieth-century visual and material culture interests in the United States, Americans had a particularly nostalgic vision of early colonial life that was conflated with the domesticity glorified in Dutch genre painting. In the nearly 50 years before the opening of the Dutch House, paintings by 'Golden Age Masters' had become ubiquitous in the United States. Collectors such as Benjamin Altman, John G. Johnson, John P. Morgan and Henry C. Frick among others influenced the aesthetic community of cities like New York and Philadelphia. The Dutch genre paintings they viewed symbolized to them Holland's democratic, republican and egalitarian ideals. In 1900 one writer proclaimed, 'Dutch art is of the people and for the people', then emphasized the vernacular, domestic subject matter of popular works: 'the home, the burghers' home not the palace, is painted'.[48] In 1889 American artist George Hitchcock published 'The Picturesque Quality of Holland Interiors and Bric-a-Brac' in *Scribner's Magazine*. Hitchcock credited '[t]he wealth, political freedom and artistic character of people' in Holland for creating 'a precious inheritance of loving artistic labour, which has filled the land with the most pleasing objects of domestic use, and made it the richest mine of the bric-a-brac hunter'.[49]

Crowninshield and her colleagues were working from a particular model well established by decades of Colonial Revival interests when they conceived how to furnish the house. In a letter to Crowninshield, a curatorial colleague noted:

> Now, that room needs only a nice little painted Dutch looking cupboard, a quaint wall sconce or two, colourful printed curtains at the window (not glazed chintz) to complete the picture. Of course, the cupboard will need a few logical utensils on it.[50]

One sales receipt from Koopman Antiques on Madison Avenue in New York itemizes some of Crowninshield's many purchases. It lists one pair of metal candlesticks for $30, one Delft vase for $25, one Hendelopen Bench for $60 and a brass bucket and copper kettle and stand for $36. These items were all to be shipped to New Castle for installation in the Dutch House.[51] By 4 December 1939, Mary W. Thompson of the DSPA wrote to Crowninshield to congratulate her: 'the table & chairs are lovely & make the little yellow room look very well'.[52] By 16 December 1939, 'six beautiful chairs had arrived' and they looked 'particularly well with the little table in the room on the right'[53] (Figure 4.4).

These six chairs are a set of turned side chairs bearing the brand of 'D-Coutong', indicating their manufacture by David Coutant, a maker active in New York City and nearby New Rochelle, New York. None of the pieces gathered by Crowninshield and her colleagues had a connection to the original inhabitants of the house. Instead, the majority of these furnishings were purchased throughout the Hudson River Valley and reflect the design and construction of that area.[54]

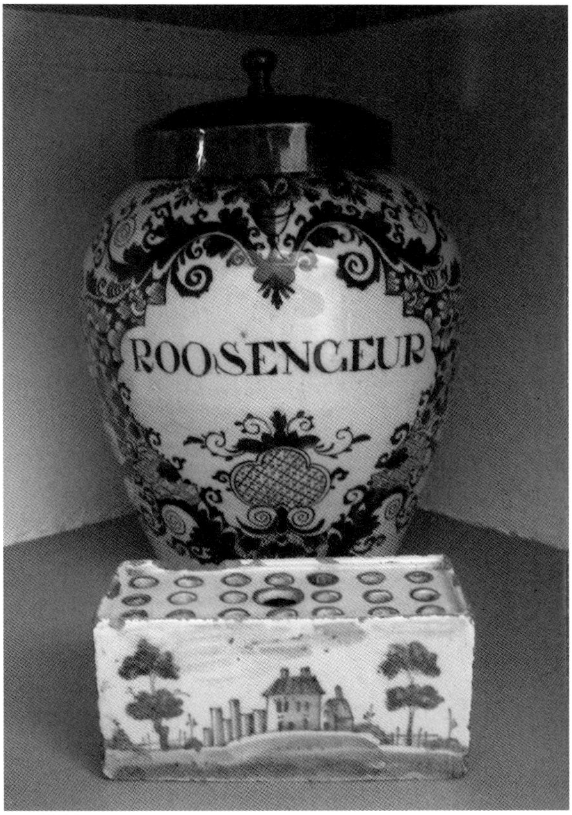

4.4 The Dutch House, Interior, 2012.

Crowninshield's role in appointing the house involved what Seth C. Bruggeman has described as 'a performance of carefully coded social rituals within a regime of highly symbolic historical objects'.[55] When faced with a lack of material contents – items from the original house were no longer in situ and the DSPA lacked a collection of works – Crowninshield filled the Dutch House with 'iconic relics', a term coined by Bruggeman to describe her curatorial selections for George Washington's birthplace. 'Indexical relics', those with a direct physical relationship to the original site, did not exist but in her search through antique dealers, auction houses and private collections throughout the North-Eastern United States, Crowninshield was able to find furnishings that reproduced closely the 'material realties' of the original house. For Bruggeman, it was 'her combined mastery of iconicity *and* her fluency in the artifactual grammar of charm that distinguished Crowninshield amid the growing male cult of authenticity'.[56]

Despite her best efforts, it now seems that many of Crowninshield's selections were more like 'symbolic relics' referencing a glorified Dutch colonial presence in New Castle. Much as in Laurel Thatcher Ulrich's study of the material culture of New England, objects now found in the Dutch House 'not only undercut colonial revival ideas about the simplicity and harmony of early American life, they challenge the compartmentalization of contemporary scholarship', in a manner that connects 'the study of household production with the study of expanding commerce'.[57] The things filling the typically Dutch interior in fact indicate the much more global reach that came to typify the worldwide expansion of the trade networks formed by the Dutch East and West India trade companies. The very goods that Crowninshield accumulated because of their 'Dutchness' were actually among some of the most 'globally' produced goods in history. The ornate rug placed on the table because it would have been too valuable to lie on the floor, was known during the late sixteenth and seventeenth century as 'Turky work', hinting at its Middle Eastern origins. While the ubiquity of these easily transportable luxury goods may have been overestimated within the visual culture of the time, they became sought-after accessories to appear in portraits that rapidly symbolized the owner's ties to international trade and cosmopolitanism. Similarly, the iconic blue and white ceramics, centrally displayed in the Dutch House and synonymous with Dutch culture, originated in Asia.[58] Prominently placed in the house were tobacco jars and flower bricks, objects that were the epitome of material transculturations in the Dutch trade empire. Through what anthropologist Alfred Kroeber called 'stimulus diffusion', Dutch culture adopted technologies and products from outside sources like Asia but gave, for example, ceramics a familiar, native content with the employment of local motifs and European functionality, drastically altering the production, circulation and consumption of visual and material culture.[59]

Historic Housekeepers: Crowninshield and her Communities

In describing the American Colonial Revival, Kenneth L. Ames, explained that it was not a surface phenomenon, but rather 'a network of communications and linkages'.[60] With the Dutch House, as with all of her projects, Crowninshield was not acting on her own: she and her brother were part of a tight network of wealthy elite collectors with similar interests. Crowninshield's collecting activities were shaped by her peers and, in particular, by her close friend Bertha Benkard, who was also instrumental in assisting du Pont at Winterthur.[61] Enabled by her family's immense wealth and this collecting network, Crowninshield occupied a privileged position within the museum world, one that was nonetheless nuanced by the gendering of American institutions. As much as she was a part of a well-documented collecting elite, Crowninshield was also a part of a community of under-represented, volunteer women, whose collaborative, behind-the-scenes efforts have been marginalized from museum history, in a manner further outlined by Anne Whitelaw's chapter in this volume (Chapter 5). Crowninshield's hard-earned success, later discounted by gendered histories, makes visible 'the line that was crossed by a few wealthy women in the late nineteenth century from being angels of culture within a domestic setting to being queens of culture in public'.[62]

Well into the twentieth century, acceptable forms of women's philanthropic involvement extended their domestic roles to encompass public concerns in a manner that has been called 'municipal housekeeping'.[63] Not-for-profit institutions and reform groups provided opportunities for women to participate in the public realm. In her work on power structures in the voluntary sphere, Kathleen D. McCarthy outlines how women's organizations were often most successful when they conformed to ideal female gender roles and arenas viewed as extensions of maternal instincts. Most importantly, McCarthy suggests that benevolent work created 'invisible' careers for women that served as a socially acceptable form of female entrepreneurship. And no project was better than the establishment of house museums, with an emphasis on the public educational roles they provided.[64]

In her work with the preservation movement, Crowninshield often negotiated very different communities – a classed one where her cultural capital provided agency, and a gendered one in which her participation was circumscribed by male domination. Like the women described by Lianne McTavish in her study of women who participated in the organization of Canada's New Brunswick Museum, Crowninshield 'garnered symbolic capital in a way that conformed to social expectations and did not overtly challenge dominant gendered norms'. Her participation placed an emphasis on domestic skills and household management in a manner in keeping with socially prescribed female roles related to caretaking.[65] Like many other women in early twentieth-century institutions, her contributions were simultaneously significant and subordinate. Along with many of her peers, Crowninshield, although self-taught rather than formally trained, gave her time and financial investments to projects, and possessed the cultural capital to become a well-respected authority. After many years in the field, prestigious heritage sites sought her opinions and input. One administrator wrote:

> it seemed wise to write to Mrs Francis Crowninshield in regard to the Wakefield work … it was not money that we needed, but interest, good judgment, and knowledge, also the ability to give *charm* to a room even with simple things. This seems to be her particular forte.[66]

In much the same way that Winterthur is touted as being formed by du Pont's solitary genius, credit for the furnishing of the Dutch House is often given solely to Crowninshield, when in fact both brother and sister operated networks of collectors, dealers, art historians, preservationists, curators and interested amateurs. Along with Thompson, others, for example Miss Sloan, cattily described in a letter as 'a rather pathetic little lady of Dutch descent', were involved in opening the house to the public. The society pages of the *Wilmington Morning News* in the summer of 1939 give names to the legion of heretofore anonymous women responsible for interpreting the house to the visiting public. While Crowninshield is the headline with her donations such as the 'Swedish Rope Bed Given to Dutch House', women including Mrs J. Leroy Holcome, Mrs Everett Reynold, Mrs Edward Taylor, Mrs R.J. Quillen and Mrs Lewis Booker were engaging with the objects in a different sort of way when they hosted tours. Furthermore, even though Crowninshield played a role in the purchase of the furnishings for the house, it is not clear how involved she was in the actual placement of objects. The director wrote that Mrs Janvier 'had not yet placed any of the things' Crowninshield had sent down. Others were involved too: when Mrs Stewart 'decided not to relinquish' her table, Mrs Bailey 'found another Delaware bench table … better in every respect. It is better looking of sturdy build, and in good condition, ready to move in, price $45'. Mrs Bailey also wished Crowninshield to consider 'a footwarmer, $5 and an old Delft vase $25 … fine and authentic'.[67] Like the spinning wheel – 'Dutch, for flax, from Maryland' – the coffee roaster, copper kettle and assorted pots, pans, spoons and forks were all acquired in order to reconstruct an Old Dutch colonial kitchen vignette.

Literally and figuratively, historic homes provided outlets for traditional women's roles. By accepting and contributing to a revisionist history of early Dutch settlements that glorified the housewife, women such as Crowninshield were able to revive women's participation in America's founding mythologies. This created a dynamic tension for those working in a public environment and fashioning private spaces: they were

> part of a wider trend in early-twentieth century middle-class women's culture, in their enjoyment of a new public role, in their defensive celebration of their 'colonial' domestic inheritance, and in their efforts to create a public institution communicating their concerns about the home.[68]

Furthermore, women's perceived affinity for decorative art objects created a unique position for their expertise in house museums. Stuart Hobbs has argued that while museum historians engaging with objects began to distinguish themselves from academic historians who traditionally privileged texts, simultaneously they 'recognized in their work a kinship with art historians who had developed methodologies for interpreting *objets d'art* just like the furniture and decorative arts displayed in house museums'.[69] Exclusion from textual-based histories provided an opportunity to study objects closely, particularly domestic objects. Participation was still gendered as a pattern emerged in house museums whereby men administrated the overall operations of the facility and women were charged with furnishing period rooms. According to West, the process of establishing a house museum 'provided a vehicle for women to negotiate the shifting relationship between women's customary power base (the home) and the public realm (the state)'.[70]

Whereas du Pont's systematic, encyclopaedic collecting can be compared to more expansive modern versions of collectors' cabinets, perhaps it is helpful to think about Crowninshield's furnishing of historic houses like dollhouses writ large. Theoretical frameworks used to interpret early modern dollhouses are equally applicable to the modern historic house preservation movement. With its open Bible, wood-stoked fire, and Crowninshield's other Dutch-like accoutrements the house would have been the epitome of historical domesticity frozen in time. In this sense, the Dutch House is one example of how the women involved in Colonial Revival and preservation communities succeeded in creating 'templates for universal order – that is dollhouses – habitable in a very real way'.[71]

According to Susan Stewart, dollhouses have two dominant motifs: that of nostalgia (clearly evidenced by Crowninshield's contribution to Colonial Revival mythologies) and that of wealth.[72] The furnishing and administration of historic houses like large-scale dollhouses by women such as Crowninshield demonstrate the masses of capital to which modern American philanthropists had access. In this sense Crowninshield's patronage becomes a socially acceptable way of displaying wealth. At a time when many Americans were struggling to survive, Crowninshield's grandiose decorative projects would have boosted regional economies. During the Depression projects with a democratically mined rhetoric were viewed as appropriate activities for a wealthy woman like Crowninshield and on the brink of a second world war nationalistically motivated, historical projects were crucial in strengthening and projecting a carefully constructed history. Second, the patriotic American image projected at the time was based on a nostalgic, colonial past. The Dutch House and its furnishings reinforced ideas of a country founded on notions of tolerance, independence and a Protestant work ethic, all viewed as initiated by Dutch settlers.

During the Depression years of the 1930s, Franklin D. Roosevelt's New Deal increased government funding to cultural institutions and over 200 more house museums would open. But with the financing came a new era of bureaucracy. In 1933, the Historic American Buildings survey was established and two years later Congress passed the Historic Sites Act in order to fund new research and establish more inventory programmes. In *Preservation Comes of Age*, Charles Hosmer considered the 1930s with the participation of major figures such as John D. Rockefeller in the preservationist movement to signal a rise in professionalism.[73] In contrast to the volunteer groups that had previously furnished sites, places such as Thomas Jefferson's Monticello and Robert E. Lee's Stratford Hall came to be curated by formally trained directors such as Fiske Kimball of what was then the Pennsylvania Museum.[74] Historic houses were taken over by men whose historic knowledge was supported by university educations. This new modernist guard of the Progressive Era decried the previous symbolic relics in favour of a new so-called authenticity that relegated the contributions of women to the margins. The legions of volunteer associations of housekeeping women were replaced with accredited men. As such, this professionalization simultaneously celebrated Crowninshield's accomplishments as a wealthy collector while it erased the collaborations that helped form many great collections. Charm was out; fact was in.

Conclusion

Published in 1898, May King Van Rensselaer's *The Goede Vrouw of Mana-Ha-Ta at Home and in Society, 1609–1760*, made clear the conflation of home and the proper Dutch–American woman's roles as homemaker. She elaborated:

> The houses of the emigrants were not bare log huts, but were at once brightened by the touch of facile feminine fingers, and converted into pleasant homes. The *geode vrouw* hung her neat lattice of leaden-sashed windows, with snowy curtains, made by her own hands, stuck a beau-pot of flowers on the ledge, set her spinning-wheel by the hearth-stone and her huge loom under the sloping roof of the 'bock stoep'.[75]

Although the Colonial Revival and the connected Holland Mania created a highly problematic sanitized view of early American life, the movement did bring a sustained interest in material culture, interiors and the preservation of historical structures.[76] Some of the stereotypical views of Dutch–American culture also, perhaps more importantly, provided a means for women to participate in the construction of the mythology that glorified domesticity. The display of 'relics' from colonial days and the recreation of vignettes, like the ubiquitous 'colonial kitchen', highlighted women's work and daily life. Increasingly, literature, art and popular culture glorified the role of the *geode vrouw* or good wife, to promote women's positions, however circumscribed by domestic boundaries, as one of agency and to a certain degree power. As genteel housekeepers, women were considered responsible for the accumulation, organization and display of material goods that, in turn, instilled in women what historian Amanda Vickery describes as a 'curatorial ethos in their dealing with things'.[77] In their role as 'mistresses' of the house, women had a profound relationship to the material goods of the domestic sphere that legitimated their role in the foundation of historic houses. As *geode vrouw*, women like Crowninshield and her colleagues legitimized their participation in the curatorship and administration of the Dutch House, through the repetition of a domestic mythology of colonial communities centred on hearth and home.

Acknowledgements

I am grateful for Janice Helland and Beverly Lemire's thoughtful comments on earlier versions of this essay. The research for this chapter was done while I was a Research Fellow at Winterthur Museum and Library. I would like to thank Rosemary Krill Winterthur for her assistance, as well as Michael Connolly, Bruce Dalleo and Judy Randy of the New Castle Historical Society.

Primary Sources Cited

Hitchcock, George, 'The Picturesque Quality of Holland Interiors and Bric-a-Brac', *Scribner's Magazine*, 5/2 (2 February 1889), pp. 162–71.

Klee, Jeff, 'The Dutch House', unpublished historical structures report.

'Louise du Pont Crowninshield Papers', Hagley Museum and Archives, Wilmington, Delaware.

'Louise du Pont Crowninshield Papers', Winterthur Museum, Garden and Library, Winterthur, Delaware.

Kruse, Gertrude and Albert, *New Castle Sketches* (Philadelphia, PA: University of Pennsylvania Press, 1932).

New Castle Deeds T-40, 263.

New Castle on the Delaware (New Castle: New Castle Historical Society, 1936), p. 71.

Rensselaer, May King Van, *The Goede Vrouw of Mana-ha-ta at Home and in Society, 1609–1760* (New York: C. Scribner's Sons, 1898).

The Collector and the Art Critic, 2 (15 February 1900), p. 137.

Secondary Sources Cited

Ames, Kenneth L., 'Introduction', in Alan Axelrod (ed.), *The Colonial Revival in America* (New York: Norton & Company), pp. 10–14.

Anderson, Benedict, *Imagined Communities: Reflections on the Origin and Spread of Nationalism* (London: Verso, 1991).

Barthel, Diane, *Historic Preservation: Collective Memory and Historical Identity* (New Brunswick, NJ: Rutgers, 1996).

Belk, Russell W. and Melanie Wallendorf, 'Of mice and men: Gender Identity in Collecting', in Susan M. Pearce (ed.), *Interpreting Objects and Collections* (London: Routledge, 1994), pp. 240–54.

Bolton, Reginald Pelham, 'The Dyckman House Park and Museum', *Beaver*, 4/2 (22 February 1917), p. 3.

Boris, Eileen, 'Crossing Boundaries: The Gendered Meaning of the Arts and Crafts', in Janet Kardon (ed.), *The Ideal Home 1900–1920* (London: Harry N. Abrams, 1993), pp. 32–45.

Bourdieu, Pierre, *Distinction: A Social Critique of the Judgment of Taste* (Cambridge: Cambridge University Press, 1984).

Burdick, Kim, 'Remembering Louise Crowninshield', *Forum Journal*, 14/3 (Spring 2000), pp. 5–6.

Broomhall, Susan, 'Imagined Domesticities in Early Modern Dutch Dollhouses', *Parergon*, 24/2 (2007), pp. 47–67.

Bruggeman, Seth C., *Here George Washington was Born: Memory, Material Culture and the Public History of a National Monument* (Athens, GA: University of Georgia Press, 2008).

Chmielewski, Laura M., 'Displaying the Dutch: The Dyckman Farmhouse Museum and the Dutch Colonial Revival', in Roger Panetta (ed.), *Dutch New York: The Roots of Hudson Valley Culture* (New York: Fordham University Press, 2009), pp. 377–92.

Corn, Wanda, *Cultural Leadership in America: Art, Matronage and Patronage* (Boston, MA: Isabella Steward Gardner Museum, 1997).

Gerard, Jessica, 'Women of the Victorian Landed Classes and the Country House', in Vanessa Dickerson (ed.), *Keeping the Victorian House* (New York: Garland Publishing, 1995), pp. 175–206.

Gordon, Beverly, *The Saturated World: Aesthetic Meaning, Intimate Objects, Women's Lives, 1890–1940* (Knoxville, TN: University of Tennessee Press, 2006).

Harper, Deborah Van Riper, '"The Gospel of New Castle": Historic Preservation in a Delaware Town', in Constance J. Cooper (ed.), *350 Years of New Castle, Delaware: Chapters in a Town's History* (Wilmington, DE: Cedar Tree Books, 2001), pp. 155–87.

Hobbs, Stuart, 'Exhibiting Antimodernism: History, Memory, and the Aestheticized Past in Mid-twentieth-century America', *The Public Historian*, 23/3 (Summer 2001), pp. 39–61.

Hobsbawm, Eric, 'Introduction: Inventing Traditions', in Eric Hobsbawm and Terence Ranger (eds), *The Invention of Tradition* (Cambridge: Cambridge University Press, 1983), pp. 1–14.

Hochstrasser, Julie, 'Re-mapping Dutch Art in Global Perspective: Other Points of View', in Mary D. Sheriff (ed.), *Cultural Contact and the Making of European Art Since the Age of Exploration* (Chapel Hill, NC: University of North Carolina Press, 2010), pp. 43–71.

Hosmer Jr, Charles B., *Preservation Comes of Age: From Williamsburg to the National Trust, 1926–1949* (Charlottesville, VA: University Press of Virginia, 1980).

Huneault, Kristina, 'Professionalism as Critical Concept and Historical Process for Women and Art in Canada', in Kristina Huneault and Janice Anderson (eds), *Rethinking Professionalism: Women and Art in Canada 1850–1970* (Montreal: McGill-Queen's University Press, 2012), pp. 3–54.

Johnson, Kathleen Eagen, 'Frans Hals to Windmills: The Arts and Crafts Fascination with the Culture of the Low Countries', in Bert Denker (ed.), *The Substance of Style: Perspectives on the American Arts and Crafts Movement* (New York: Norton & Company, 1990), pp. 47–67.

Kroeber, Alfred, *Anthropology: Culture Patterns and Processes* (New York: Harcourt, Brace & World, 1963).

Maynard, W. Barksdale, 'New Castle's Dutch Tile House of 1687: Fraud or Genuine?' *Delaware History*, 29 (2001), pp. 141–68.

McCarthy, Kathleen D., *Lady Bountiful Revisited: Women, Philanthropy and Power* (London: Rutgers, 1990).

McNeil, Peter, 'Crafting queer spaces: Privacy and posturing', in Alla Myzelev and John Potvin (eds), *Fashion, Interior Design and the Contours of Modern Identity* (Burlington, VT and Aldershot: Ashgate, 2010), pp. 19–41.

McTavish, Lianne, 'The Rewards of Professionalization: Alice Lusk Webster and the New Brunswick Museum, 1933–53', in Kristina Huneault and Janice Anderson (eds), *Rethinking Professionalism: Women and Art in Canada 1850–1970* (Montreal: McGill-Queen's University Press, 2012), pp. 83–104.

New Castle Historical Society, 'The Dutch House'. Available at: http://www.newcastlehistory.org/houses/collections.html [accessed 3 February 2012].

Rybczynski, Witold, *Home: A Short History of an Idea* (New York: Viking, 1986).

Stewart, Susan, *On Longing: Narratives of the Miniature, the Gigantic, the Souvenir, the Collection* (Durham, NC: Duke University Press, 1993).

Stillinger, Elizabeth, *The Antiquers* (New York: Alfred A. Knopf, 1980).

Stott, Annette, *Holland Mania: The Unknown Dutch Period in American Art and Culture* (Woodstock: Overlook Press, 1998).

Terry, Andrea, 'Gender, Canadian Nationhood and "Keeping House": The Cultural Bureaucratization of Dundurn Castle in Hamilton Ontario, 1900–1960s', *Gender & History*, 25/1 (April 2013), pp. 47–63.

Ulrich, Laurel Thatcher, *The Age of Homespun: Objects and Stories in the Creation of an American Myth* (New York: Knopf, 2001).

Vickery, Amanda, 'Women and the world of goods: a Lancashire consumer and her possessions, 1751–81', in John Brewer and Roy Porter (eds), *Consumption and the World of Goods* (New York: Routledge, 1993), pp. 274–303.

Vincentelli, Moira, *Women and Ceramics: Gendered Vessels* (Manchester: Manchester University Press, 2000).

West, Patricia, *Domesticating History: The Political Origins of America's House Museums* (Washington, DC: Smithsonian Institution Press, 1999).

Wilson, Richard Guy, 'What is the Colonial Revival?' in Richard Guy Wilson, Shaun Eyring and Kenny Marotta (eds), *Re-creating the American Past: Essays on the Colonial Revival* (Charlottesville, VA: University of Virginia Press, 2006), pp. 1–12.

Zimmerman, Philip D., 'Early American furniture in the New Castle Historical Society in Delaware', *The Magazine Antiques* (May 2005), pp. 130–41.

Notes

1 Gertrude and Albert Kruse, *New Castle Sketches* (Philadelphia, PA, 1932), n.p.

2 Seth C. Bruggeman, *Here George Washington was Born: Memory, Material Culture and the Public History of a National Monument* (Athens, GA, 2008), p. 130.

3 'Mrs Crowninshield Tells of Restoration Problems', Author unknown, undated article. Hagley Archives, Crowninshield Papers 471.

4 Diane Barthel, *Historic Preservation: Collective Memory and Historical Identity* (New Brunswick, NJ, 1996), p. 36. Original emphasis included.

5 Laurel Thatcher Ulrich, *The Age of Homespun: Objects and Stories in the Creation of an American Myth* (New York, 2001).

6 *New Castle on the Delaware* (New Castle, DE, 1936), p. 71.

7 Annette Stott, *Holland Mania: The Unknown Dutch Period in American Art and Culture* (Woodstock, 1998), pp. 191.

8 New Castle Deeds T-40, 263.

9 Jeff Klee, 'The Dutch House', unpublished historical structures report.

10 Ibid.

11 Deborah Van Riper Harper, '"The Gospel of New Castle": Historic Preservation in a Delaware Town', in Constance J. Cooper (ed.), *350 Years of New Castle, Delaware: Chapters in a Town's History* (Wilmington, DE, 2001), p. 158.

12 W. Barksdale Maynard, 'New Castle's Dutch Tile House of 1687: Fraud or Genuine?', *Delaware History*, 29 (2001), pp. 141–68.

13 Harper, '"The Gospel of New Castle"', p. 187.

14 Ibid., pp. 166–7.

15 Bruggeman, *Here George Washington was Born*, p. 21.

16 As part of her application to The National Society of The Colonial Dames of America In The State of Delaware, Crowninshield's mother Evelina traced her family back to John Bleeker, who was born in Meppel, Holland, but came to New Amsterdam in 1658 and resided there until his death in 1732. Furthering a Dutch connection, her paternal grandmother was a member of the prestigious Albany Ten Eyck family. Hagley Archives, 471, Box 37.

17 In 1897 Crowninshield's founded 'Willing Helpers', a group of young women who met Saturday mornings to make clothing for infants at a Wilmington day nursery. Her charitable participation

was not only monetary, for she was volunteered at the Marblehead Children's Island and the Boston Lying-In Hospital.

18 In 1955, President Dwight D. Eisenhower appointed her to the Boston National Historic Sites Commission. She was also involved with historic restoration of the Kenmore Association for the home of Fielding Lewis and Betty Washington, George Washington's sister, and many other projects in Massachusetts including Gore Place, Waltham; Lee Mansion, Marblehead; Wayside Inn, Sudbury; Pierce-Nichols House, Derby House and Gardner-Pingree House, Salem; Mission House, Stockbridge. In Philadelphia, PA she helped establish the Princeton Battle Park and the helped in the restoration of Independence Hall and Society Hill. Trustee of the Peabody Museum and Essex Institute, Salem, MA; president of the Salem Maritime Trust; board member of the Boston National Historic Sites Commission; Director of the First Iron Works, Saugus, MA. Like her brother, she was also an avid horticulturalist, serving as director of the Arnold Arboretum, Boston, MA and vice president of the Garden Club of America. Crowninshield also helped Truman's White House Redecorating committee.

19 Undated speech. Hagley Archives. In recognition of her services to historic preservation, the National Trust instituted an annual award, the Louise Evelina du Pont Crowninshield Award, for achievement in the preservation and interpretation of historic, architectural and maritime heritage.

20 Kristina Huneault, 'Professionalism as Critical Concept and Historical Process for Women and Art in Canada', in Kristina Huneault and Janice Anderson (eds), *Rethinking Professionalism: Women and Art in Canada 1850–1970* (Montreal, 2012), pp. 42–4.

21 See http://www.preservationnation.org/forum/library/public-articles/remembering-louise-crowninshield.html [accessed 23 October 2013].

22 Susan Broomhall, 'Imagined Domesticities in Early Modern Dutch Dollhouses', *Parergon*, 24/2 (2007), p. 47.

23 Russell W. Belk and Melanie Wallendorf, 'Of mice and men: Gender Identity in Collecting', in Susan M. Pearce (ed.), *Interpreting Objects and Collections* (London, 1994), p. 240.

24 Pierre Bourdieu, *Distinction: A Social Critique of the Judgment of Taste* (Cambridge, 1984).

25 Wanda Corn, 'Art Matronage in Post-Victorian America', *Cultural Leadership in America: Art, Matronage and Patronage* (Boston, MA, 1997), p. 14.

26 One study characterizes Henry Francis du Pont's collecting activities as creating a 'innovative "queer space" for the projection of fluid male identities and fantasies employing intimate private spaces, furnishings, dress and diversions'. Peter McNeil, 'Crafting queer spaces: privacy and posturing', in Alla Myzelev and John Potvin (eds), *Fashion, Interior Design and the Contours of Modern Identity* (Burlington, VT and Aldershot, 2010), p. 19.

27 Moira Vincentelli, *Women and Ceramics: Gendered Vessels* (Manchester, 2000), pp. 109–10.

28 Elizabeth Stillinger, *The Antiquers* (New York, 1980), pp. 245–8. While Gordon finds this categorization problematic, I have slightly different concerns. In this book other well-respected male collectors – George Francis Dow, Henry Ford and even Henry Francis du Pont – are each given chapters. My greater concern is how Crowninshield is appended with two other female collectors – Katharine Prentis Murphy and Electra Havemeyer Webb – in a chapter entitled 'A Triumvirate and One Lone Texan', almost as an afterthought. Beverly Gordon, *The Saturated World: Aesthetic Meaning, Intimate Objects, Women's Lives, 1890–1940* (Knoxville, TN, 2006).

29 Jessica Gerard, 'Women of the Victorian Landed Classes and the Country House', in Vanessa D. Dickerson (ed.), *Keeping the Victorian House* (New York, 1995), p. 176.

30 Gordon, *The Saturated World*, pp. 194–6.

31 The New Castle Historical Society, 'The Dutch House'. Available at: http://www.newcastlehistory.org/houses/collections.html [accessed 6 September 2012].

32 Stott, *Holland Mania.*

33 Kathleen Eagen Johnson, 'Frans Hals to Windmills: The Arts and Crafts Fascination with the Culture of the Low Countries', in Bert Denker (ed.), *The Substance of Style: Perspectives on the American Arts and Crafts Movement* (New York, 1990), p. 47.

34 Letter from Douglas Campbell to Louise Crowninshield, Hagley Archives 471.

35 Benedict Anderson, *Imagined Communities: Reflections on the Origin and Spread of Nationalism* (London, 1991).

36 Richard Guy Wilson, 'What is the Colonial Revival?' in Richard Guy Wilson, Shaun Eyring and Kenny Marotta (eds), *Re-creating the American Past: Essays on the Colonial Revival* (Charlottesville, VA, 2006), p. 3.

37 Johnson, 'Frans Hals to Windmills', p. 52.

38 Quoted in Stott, *Holland Mania*, p. 95.

39 Reginald Pelham Bolton, 'The Dyckman House Park and Museum', *Beaver*, 4/2 (February 22, 1917), p. 3.

40 Johnson, 'Frans Hals to Windmills', p. 48.

41 Ibid., p. 47.

42 Witold Rybczynski, *Home: A Short History of an Idea* (New York, 1986).

43 Stott, *Holland Mania*, p. 186.

44 Bruggeman, *Here George Washington was Born*, p. 62.

45 Eric Hobsbawm, 'Introduction: Inventing Traditions', in Eric Hobsbawm and Terence Ranger (eds), *The Invention of Tradition* (Cambridge, 1983), p. 1.

46 Bruggeman, *Here George Washington was Born*, p. 116.

47 The New Castle Historical Society, 'The Dutch House'. Available at: http://www.newcastlehistory.org/houses/collections.html [accessed 19 April 2012].

48 Anonymous, *The Collector and the Art Critic*, 2 (15 February 1900), p. 137.

49 George Hitchcock, 'The Picturesque Quality of Holland Interiors and Bric-a-Brac', *Scribner's Magazine*, 5/2 (2 February 1889), p. 162.

50 Winterthur Museum and Archives.

51 Hagley Museum and Archives, 471, Box 54.

52 Winterthur Museum and Archives.

53 Winterthur Museum and Archives.

54 Philip D. Zimmerman, 'Early American furniture in the New Castle Historical Society in Delaware', *The Magazine Antiques* (May 2005), pp. 130–41.

55 Bruggeman, *Here George Washington was Born*, p. 63.

56 Ibid., p. 116.

57 Urich, *The Age of Homespun*, p. 7.

58 Julie Hochstrasser, 'Re-mapping Dutch Art in Global Perspective: Other Points of View', in Mary D. Sheriff (ed.), *Cultural Contact and the Making of European Art Since the Age of Exploration* (Chapel Hill, NC, 2010), pp. 43–71.

59 Alfred Kroeber, *Anthropology: Culture Patterns and Processes* (New York, 1963), pp. 176–8.

60 Kenneth L. Ames, 'Introduction', in Alan Axelrod (ed.), *The Colonial Revival in America* (New York, 1985), p. 3.

61 Along with filling a multitude of positions in cultural organizations, Benkard was president general of the Colonial Dames of America and was considered an expert on eighteenth-century furniture, and Duncan Phyfe's work specifically. Through Benkard, Crowninshield became involved in numerous projects in the 1930s including Kenmore, the home of George Washington's sister Betty and her husband, Fielding Lewis.

62 Corn, *Cultural Leadership in America*, p. 23.

63 Eileen Boris, 'Crossing Boundaries: The Gendered Meaning of the Arts and Crafts', in Janet Kardon (ed.), *The Ideal Home 1900–1920* (London, 1993), p. 34.

64 Kathleen D. McCarthy, *Lady Bountiful Revisited: Women, Philanthropy and Power* (London, 1990), pp. 1–10.

65 Lianne McTavish, 'The Rewards of Professionalization: Alice Lusk Webster and the New Brunswick Museum, 1933–53', in Kristina Huneault and Janice Anderson (eds), *Rethinking Professionalism: Women and Art in Canada 1850–1970* (Montreal, 2012), p. 89.

66 Bruggeman, *Here George Washington was Born*, p. 129.

67 Winterthur Archives.

68 West, *Domesticating History*, p. 72.

69 Stuart Hobbs, 'Exhibiting Antimodernism: History, Memory, and the Aestheticized Past in Mid-twentieth-century America', *The Public Historian*, 23/3 (Summer 2001), p. 55.

70 Andrea Terry, 'Gender, Canadian Nationhood and "Keeping House": The Cultural Bureaucratisation of Dundurn Castle in Hamilton, Ontario, 1900–1960s', *Gender & History*, 25/1 (April 2013), p. 48.

71 Bruggeman, *Here George Washington was Born*, p. 60.

72 Susan Stewart, *On Longing: Narratives of the Miniature, the Gigantic, the Souvenir, the Collection* (Durham, NC, 1993), p. 61.

73 Charles B. Hosmer Jr, *Preservation Comes of Age: From Williamsburg to the National Trust, 1926–1949* (Charlottesville, VA, 1980).

74 Bruggeman, *Here George Washington was Born*, p. 127.

75 May King Van Rensselaer, *The Goede Vrouw of Mana-ha-ta at Home and in Society, 1609–1760* (New York, 1898).

76 Laura M. Chmielewski, 'Displaying the Dutch: The Dyckman Farmhouse Museum and the Dutch Colonial Revival', in Roger Panetta (ed.), *Dutch New York: The Roots of Hudson Valley Culture* (New York, 2009), p. 387.

77 Amanda Vickery, 'Women and the world of goods: a Lancashire consumer and her possessions, 1751–81', in John Brewer and Roy Porter (eds), *Consumption and the World of Goods* (New York, 1993), p. 282.

5

From the Gift Shop to the Permanent Collection: Women and the Circulation of Inuit Art

Anne Whitelaw

At a meeting in June 1969, the members of the Women's Society of the Edmonton Art Gallery voted unanimously to spend a $2,000 savings bond bought to mark Canada's centennial on the purchase of 'Eskimo[1] sculptures' for donation to the gallery. Although the museum's women's society had been in existence since 1944, and was the initiator of many of the gallery's social and fundraising activities, this was its first direct purchase of works of art for the permanent collection. This lack of participation in decisions regarding the acquisitions enabled by its funds was not unusual. With the significant exception of the Women's Committee of the Art Gallery of Ontario, which made important acquisitions of contemporary art, volunteer women's societies in art museums across the country had virtually no say in which works, or even what kinds of works, were purchased with the monies they raised. Inuit art, however, seems to have been one collecting area over which women's societies had more agency. Introduced to the broad Canadian public in the late 1940s, Inuit sculptures would have been familiar to many of these women through their presence in members' homes, or through their sale in the galleries' gift shops – operations founded and run by the same volunteer women's societies. Inuit art enjoyed great popular success throughout the 1950s and 1960s, and with the introduction of the annual print collections in 1960 was rapidly taken on by art dealers and some commercial galleries. Public art museums, however, were slower to welcome Inuit art into their permanent collections because of what was still considered to be its primitive character and the belief that such work was better suited to the collections of ethnographic museums.

Today, Inuit art is a central component of almost every major art museum in Canada and there is little question that prints, sculptures and work in a variety of other media produced by artists of Inuit descent have as clear a place in the history of Canadian art as the work of any southern-born artist. But while the aesthetic status of Inuit art or its place in the art museum is no longer in question, how it came to be included in the institution is less known. As the above anecdote suggests, volunteer women's groups played a major role in bringing Inuit art into the museum and making it part of permanent collections, but this work has been overlooked in the literature, often in favour of the significant donations of major (male) collectors. While these donations were and continue to be extraordinarily important in solidifying the presence of Inuit sculpture and prints in art museums and in increasing these museums' holdings of such objects, the work of women in facilitating the introduction of Inuit art into the museum warrants examination. Such an investigation will not only fill in the gaps relating to early acquisitions of Inuit sculpture and prints, but it will also raise awareness of the central role of women – particularly voluntary women's groups – in bringing Inuit art from museums' gift shops into their permanent collections. This essay will also argue that the key to understanding the success of women's committees in bringing Inuit art into museums is to recognize how these volunteer organizations functioned for the women themselves. I view women's committees as spaces of sociability that gave volunteer women some standing within the art museum, providing them with membership in a community of like-minded women, and giving a focus to their energy and intelligence. Most active in the decades following the Second World War, women's committees organized a significant range of activities for North American art museums, which have since become vital components of these institutions' programming. Regular committee meetings and the organization of major projects created a powerful sense of belonging and purpose for these women, whose class, race and marital status obviated the necessity for them to take on paid work. The efforts of these groups, however, should not be dismissed as dilettantish; rather, members of women's committees worked hard and undertook a broad range of tasks that would be more commonly associated with paid labour and, by the 1980s, had become professionalized services within the museum itself.

The Development of Inuit Art

The development of Inuit art as a commodity designed to improve the economic conditions of the Indigenous inhabitants of the Arctic dates from the late 1920s and involves three main players: the federal government, the Canadian Handicrafts Guild (CHG) and the Hudson's Bay Company (HBC).[2] The federal government, through the auspices of the Department of the Interior, first noted in 1928 that the production of handicrafts for sale in the South might alleviate some of the economic hardships faced by Inuit communities when the fox population was low and revenue from trapping was down. While little came of this first proposal, what had by then become the Department of Mines and Resources revived the proposal in the late 1930s with the particular aim of assisting old and infirm inhabitants of the industrial homes and hospitals in Chesterfield Inlet and Pangnirtung. Meanwhile, the women of the CHG in Montreal, already encouraging means of economic support to ethnic and First Nations communities through the sale of handicrafts, were aware of the ivory carvings and sealskin objects made by the Inuit and had included some examples of these among the goods for sale in the Guild's handicrafts store. But it was through contact with James Houston, a young artist from Grand-Mère Quebec who had travelled to Port Harrison (now Inukjuak) to paint and had returned with a small selection of carvings produced by Inuit hunters, that the Guild planned a systematic programme of assistance in the Arctic. This programme was significantly boosted by a formal arrangement with the Department of Mines and Resources (later the Department of Northern Affairs and National Resources) to develop craft production as a means of economic subsistence in the North. This entailed contracting Houston to encourage carving among the communities growing around the various HBC posts, buying the resulting objects, and bringing them back to Montreal for sale in the Guild's shop.

As the primary commercial outlet in the Arctic, the HBC was a crucial participant in the establishment of a national market for Inuit handicrafts. Its trading posts and stores were the sites where artists could exchange their work for money or store credit; more importantly, the company's extensive shipping system ensured the transportation of what would soon become vast numbers of cases of carvings from Arctic Canada to the South, along with the personnel (such as Houston) who would foster the developing carving programme. As the primary location in the South for the sale of Inuit carvings, the HBC continued to exercise an important role in the circulation and consumption of Inuit handicrafts. Carvings were available in virtually all major HBC stores across the country as part of their 'Canadiana' section, which specialized in souvenirs. The scale of the HBC's retail operation worried the members of the CHG, who were concerned that the potential profits generated by the sale of carvings would spur the company to emphasize commercial rather than aesthetic qualities in the carvings they selected for purchase. However, the Guild also realized that, as a small voluntary organization, it could not finance what was becoming a significant market for Inuit carvings.

As this overly simplified account suggests, the federal government, the HBC and the CHG worked jointly in the production of the industry now known as 'Inuit art'. The relationship between these three players as well as their differing motivations for encouraging the trade in Inuit art warrant examination for the insight such interactions provide not only into the development of Inuit art as an art form, but also into the manner in which the market for Inuit art was shaped. For the federal government, as we have seen, encouraging a market for Inuit carving was a way of increasing self-sufficiency among the Inuit and reducing the need for welfare payments. While they occasionally balked at the costs of setting up the programme (such as paying for Houston's trips), government staff quickly recognized the value of encouraging what were essentially make-work programmes once they saw the success of the product across the country. For the HBC, access to a commodity with such clear market value was a key motivator for its continued participation in the programme. Indeed, when the CHG was no longer capable of underwriting the purchase of carvings from the artists, the HBC stepped in, thereby increasing its participation from transporting sculptures and personnel to purchasing all saleable carvings produced in the North for eventual distribution to their own retail stores and to the Guild for sale.

As the organization with the most experience in the promotion of craft to a national public and with a longstanding commitment to supporting traditional Indigenous forms, the CHG played a key role in maintaining quality control over the carvings that were made available for sale. The Guild's Indian and Eskimo Committee under the leadership of Alice Lighthall as well as the Shop Committee chaired by C.J.G. Molson had very clear ideas about what kinds of carvings would sell. But their greater concern lay with ensuring that Inuit were encouraged to produce the best work possible, rather than churning out objects that would be considered commercially viable. Not only through what was on sale in its own shops, but also in the objects shipped to select stores and groups across the country, the Guild was the main purveyor of the carvings that had captured the public's imagination, and it maintained a close eye on the quality of the carvings offered for sale. The Guild also monitored what outlets other than its handicrafts stores would be given the opportunity to offer this hot new commodity for sale. Even in relation to the carvings sold by the HBC, the CHG maintained a certain degree of control, regularly cautioning representatives of the HBC and government officials that the primary requirement in the purchase of Inuit carvings was their authenticity and the skill of the workmanship. Quality and authenticity were important criteria for the Guild and from the beginning of the project, its members voiced strong concerns that leaving marketing and sales decisions in the hands of the HBC could result in over-commercialization and a focus on profit share rather than the welfare of the artist. The Guild had expressed concern that this attitude also underpinned the government's initial interest in Inuit carving, but, as can be seen in this 1955 letter to the Guild's C.J.G. Molson, the Department of Northern Affairs and National Resources shared the Guild's concern that too great an emphasis on commerce would have a detrimental effect on the aesthetic form as well as the long-term economic sustainability of the project for the Inuit:

> As you may know, we have been giving considerable thought to arts and handicrafts in relation to the problems of the Eskimos. Like you, we are anxious above all that carving and similar activities should be carried on to the long-term advantage of the Eskimos themselves. We also attach great importance to the maintenance of high standards in Eskimo art. There are, however, pitfalls and dangers. We are much aware of the debasement of native arts and handicrafts in other parts of the world from which we have had reports, and we are particularly anxious that similar misfortunes should not occur in the Canadian north.[3]

As public interest in 'Eskimo art' increased dramatically with the publication of articles and photo essays in both art and general-interest magazines – in 1954 alone, articles appeared in *Canadian Art*, *The Studio* and *The Beaver*, while photo essays were published in May of that year in both *Life* and *Vogue* – the need to closely monitor the commercialization of Inuit carving had to be balanced against the high rate of sales these objects commanded. Stories of CHG exhibitions selling out within days of opening were frequently repeated by Guild members as a way of promoting the new form; and as will be discussed below, Guild shop staff were keenly aware of what kinds of carvings sold well and they encouraged Arctic carvers to produce work predominantly in that vein. Over-commercialization was associated more closely with department stores and unscrupulous dealers (as well as short-sighted government officials) seeking to profit from a newly introduced commodity. For the CHG, offering Inuit carvings for sale was instrumental in preserving cultural values and creating opportunities for economic subsistence to a people facing destitution. Indeed, the Guild's Indian and Eskimo Committee articulated its mandate as 'the encouragement of the best use of traditional design, material and workmanship and the education of an appreciative and discerning public',[4] and it realized these aims through the buying and selling of quality goods.

Women and the Selling of Inuit Art

In its consideration of the work of members of the CHG, the above account acknowledges the fundamentally formative role of women in the production of this cultural form. The work of the Guild is particularly important: most – although not all – of its members were female,[5] and they viewed women as the primary conduits of information about handcraft to Inuit artists. In 1948, the Guild sent letters and a leaflet with 'Suggestions for Eskimo Handicrafts' to the wives of HBC employees, as well as to female missionaries, teachers and nuns who might be working in the Arctic and who could provide instruction to the Inuit on 'appropriate' craft production.[6] And women continued to be seen to play an important role in the development of art production in the North: Helga Goetz has reported on Department of Northern Affairs discussions in the 1960s regarding the employment of the wives of men stationed in the Arctic to facilitate the production of handicrafts. It concluded that 'the average wife could be expected to devote twenty five hours of her time per month to "volunteer" work in the community',[7] but that she should be paid for any additional work thereafter. The view that married women's work would be voluntary first and paid (if at all) second is an assumption about the value of women's labour that also underscores attitudes towards volunteer women's societies in art museums.

The role of women in the circulation of Inuit art, however, extended beyond the work of the CHG and highlights how volunteer committees constituted an important and effective cross-country network. Women working in branches of the Guild across Canada were pivotal in bringing Inuit art to the attention of the larger public by making carvings available for sale in their main Montreal shop and in the stores and sales organized by Guild branches in other cities, including Edmonton (Figure 5.1). The Guild was also the main point of contact for other organizations seeking to offer Inuit carvings for sale in local handicraft sales or Christmas bazaars. In 1954, for example, Mrs H.A. Cave of the Women's Committee of the Windsor Art Association wrote to Martha Peters of the CHG asking if the Association could have some carvings for its annual Christmas Sale of Canadian Crafts. Seeking objects in the eight-, ten- and fifteen-dollar range, Cave noted that there was great interest in the sculptures and that she was already fielding questions about whether there would be any in the Christmas Sale: 'We certainly were successful in selling the last lot, and hope this year to sell all you can send us'.[8] The CHG fielded numerous such requests and was happy to fill them so long as stock was available and the requesting organization was viewed as an appropriate venue for the presentation and sale of Inuit art. However, it insisted that shipment of carvings to other organizations had to wait until the Guild had concluded its own sale of new carvings in early November. This controlled circulation of Inuit art ensured that

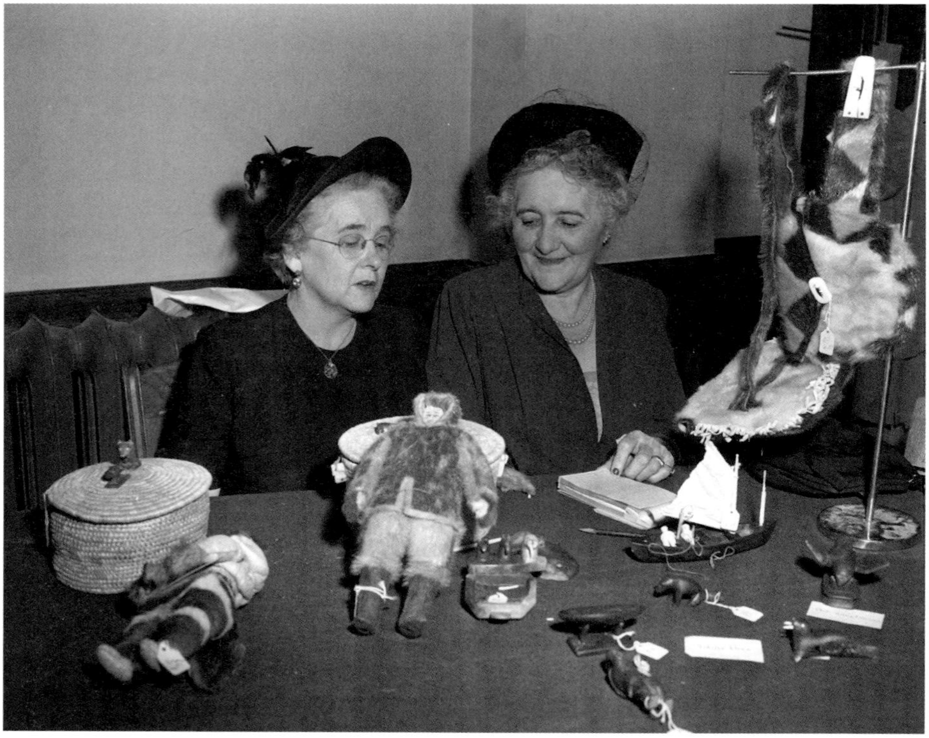

5.1 'Women's Handicraft Guild members Mrs. J. Strachan (left) and Mrs. J. Harley display Eskimo handiworks', from the Edmonton *Bulletin*, November 25, 1950.

the Guild maintained its position as the primary purveyor of quality carvings to the public through its own handicrafts store, as well as in its careful dissemination of works to deserving secondary outlets. While this vigilance could not extend to the objects sold by the HBC – or to the manner in which HBC stores presented this work – the Guild actively restricted access to Inuit sculpture to those retail outlets it felt provided a high-quality sales environment. Retailer Bowring Brothers, for example, was given approval to sell Inuit sculpture in its store in the newly opened airport in Montreal after the success of its shop in the airport at Gander, but was forbidden from selling any carvings in its downtown store.[9] Similarly, upon hearing that carvings were available 'all over' Toronto, members of the CHG asked their contacts in that city to investigate, and wrote several strongly worded letters to both the HBC and the Department of Northern Affairs reinforcing the importance of keeping the sale of Inuit art from becoming a crassly commercial venture.[10]

While much of the appeal of Inuit sculpture was its exoticism and, as Kristin Potter has argued, the possibilities it afforded for armchair tourism,[11] another reason for the successful sales of carvings from the moment it was introduced in the South was its price: a chart from 1965 indicating the most popular price points for the sale of Inuit sculpture clearly shows that the majority of purchases were made in the one-to-eight dollar range, a price that made carvings accessible to almost any buyer[12] (Figure 5.2).

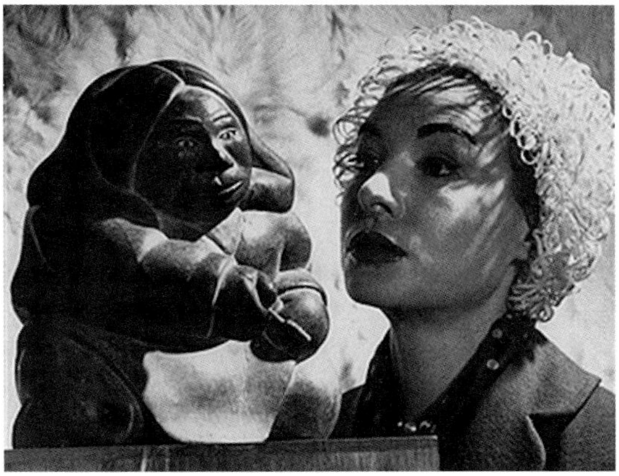

5.2 Photograph of Mrs Robert Miller, from the *Life* magazine article 'Church Sale of Arctic Art', May 25, 1959, p. 43.

It may have been the price that initially encouraged museum gift shops to request samples of Inuit carvings to sell to their customers. In September 1954, four months after articles were published in both *Vogue* and *Life* magazines, Carl Fox, manager of The Gallery Shop at the Brooklyn Museum of Art, wrote to the CHG saying that he was 'searching for authentic native handicrafts to feature and sell to our visitors'. He continued:

> So far we have been fortunate in finding some unusual things from Sweden, Japan, Mexico, South America and Africa, but sadly lack pieces from the States and our neighbors to the North. In the main, my purchases have been directed toward supplying our children visitors with representative folk arts and toys of other countries.[13]

As this anecdote reveals, 'Eskimo art' was one of several 'native' or 'primitive' cultural forms that served the curiosity market of the time. And while certain forms of Indigenous cultural production were occasionally displayed so as to reinforce formal connections to modern art – as can be seen, for example, in the Indian Art of the United States exhibition at the Museum of Modern Art in 1941[14] – such presentations were the exceptions to the dominant view that firmly placed such cultural production within the category of ethnographic artefact or exotic souvenir. Thus, while Eskimo handcraft was more than welcome in the Brooklyn Museum of Art's gift shop, it was not going anywhere near its permanent collection.

In the 1950s, however, gift shops were relatively new additions to museums. As the example of the Gallery Shop at the Brooklyn Museum suggests, American institutions were much quicker to realize the money-making as well as marketing possibilities that accompanied the presence of a retail outlet. Such major museums as the Metropolitan Museum of Art and the Museum of Modern Art in New York had well-established museum shops by the 1930s that sold print reproductions of highlights of the museums' collections as well as an assortment of other tasteful art-related objects.[15] In Canada, the National Gallery initiated a series of postcard-size reproductions of major works from its Canadian collection in the 1930s, which it not only sold at its information desk but also made available to other museums and institutions across the country. A stand-alone gift shop, however, was much slower in coming to the National Gallery and to other institutions across the country.

As with most ancillary services in Canadian museums, gift shops were introduced and staffed by members of the museums' voluntary women's societies. Formed in the mid-1940s to raise awareness of the museum as well as to raise funds, women's societies initiated and ran many activities and programmes, including: organizing publicity for museum events; preparing and serving refreshments at exhibition openings; hosting membership teas or luncheons for prospective members; running the art sales and rental galleries; organizing sales of contemporary art; running docent and other education programmes; and generally doing much of the day-to-day work that has since been turned over to paid full-time staff in the contemporary institution.[16] Many women's committees organized yearly Christmas bazaars or rummage sales throughout their history, but a permanent gift shop was a heavy commitment for a voluntary organization. The Art Gallery of Ontario (AGO)'s Women's Committee's deliberations over starting a gift shop are telling in this regard. First proposed in the fall of 1953 as a source of continuous revenue for the Women's Committee's purchase fund, the Gallery Shop's aim was 'to utilize the taste, discrimination and freedom of travel enjoyed by the members of our committee' and to offer 'only approved articles for sale, of an unusual, artistic, and personalized nature'.[17] Some members voiced concerns that the Gallery Shop might be undignified and 'too commercial and revolutionary for an Art Gallery'.[18] However, the projected financial benefits outweighed the potential indignity to the museum and the Gallery Shop opened for the 1953 Christmas season and went on to raise significant amounts of money for the AGO's acquisition fund.

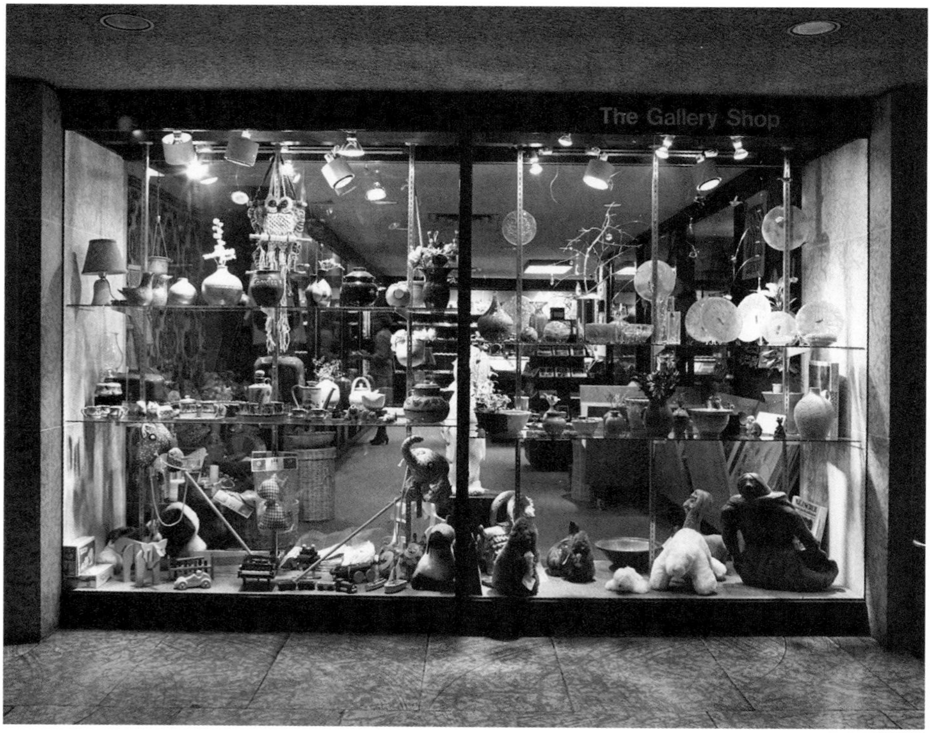

5.3 The Gift Shop at the Winnipeg Art Gallery, 1979.

Other Canadian institutions slowly followed over the course of the 1950s and 1960s, with their women's societies establishing full gift shops or even simply setting up locations for the sale of reproductions and catalogues. At the Edmonton Art Gallery, the Women's Society had established a shop as early as 1945. 'The Museum Shop' was formed just one year after the society's own founding, and was conceived as both a money-making operation for the society and what that year's annual report described as a 'much needed outlet for the better handicraft of the community'.[19] There is little mention of The Museum Shop in the Edmonton Art Gallery Women's Society's files from the following decade, but references to it appear regularly in the minutes of the Edmonton Branch of the CHG, where it is described as one of two outlets in the province to handle handicrafts exclusively – and therefore an ideal location for the sale of Inuit carvings. The lack of discussion of The Museum Shop within both the Women's Society's files and the minutes of the museum's Board of Trustees suggests that the shop was small in scale and successful enough not to be an area of concern. Among the handicrafts sold by The Museum Shop were objects obtained from the Grenfell Mission in Newfoundland/Labrador and from patients at the Camsell Hospital in Edmonton.[20] Objects from both locations included what were described as 'Eskimo crafts'[21] and were popular sellers. Other items available at The Museum Shop included National Gallery-issued art reproductions and 'hasti-notes' based on designs by local artists. During the late 1950s, however, these items were presented as being available at the Art Gallery's information desk, rather than at the kind of dedicated space that the title 'The Museum Shop' might suggest. Such a stand-alone space would not become available until the opening of the Edmonton Art Gallery's new building in 1969.

'Selling' Inuit Art to the Public Art Gallery

While carvings were available for purchase in art museum gift shops (Figure 5.3), it was with the introduction of printmaking that Inuit art's acceptance by the art museum became more evident. Initiated by James Houston[22] with the full backing of the Department of Northern Affairs and National Resources as another programme designed to offer a means of economic subsistence to impoverished Inuit communities, the introduction of printmaking was also seen as a way of maintaining the supply of Inuit art during occasional dips in the availability of soapstone.[23] Already skilled in carving, Inuit artists adapted easily to printmaking and by the summer of 1959, one year after the programme was launched in Cape Dorset, a small exhibition of prints went on display at the Stratford Festival with pamphlets indicating that editions of the prints would be on sale at art museums and galleries across the country in the fall.[24] And indeed, many art museums – through the actions of their women's committees – began to sell Inuit prints in their gift shops. As just one example, the Women's Society of the Edmonton Art Gallery announced in its 1959–60 annual report that it had taken on a new venture that year: the sale of 'Eskimo Prints'. Starting with the purchase of 20 prints in May 1960 to gauge public interest, these and a selection of Ganymede prints were offered for sale at the gallery's monthly 'Art Rental' nights.[25] The success of the venture is somewhat difficult to gauge as the Women's Society minutes report the decision on 14 March 1963 to 'take on the Eskimo prints as a project *again*'.[26] Buying a set of 17 prints from the edition that was released for sale across the country on 17 April, the renewed venture was launched at a coffee party on 30 April that also featured a lecture by then Gallery Director John MacGillivray. Unfortunately, sales of the prints were not strong in Edmonton, which may have had something to do with the lack of proper shop facilities at the museum, and competition from the other works available at the Art Rental evenings. In contrast, the 1962–63 annual report of the Vancouver Art Gallery's Women's Auxiliary noted strong sales of both prints and carvings along with the assertion that the Gallery shop was the only outlet for Inuit art in the city, a situation that would certainly change in the years to come.

The decision to sell prints in art museums and galleries was strategic on the part of the Department of Northern Affairs and National Resources, as it created a clear association between the prints and museums' status as arbiters of aesthetic quality. Even more tellingly, the decision was made specifically *not* to make the prints available for sale through the HBC as it was felt that the association with souvenirs available in the department store would diminish the prints' value as 'fine art'. And indeed, the sale of what have come to be known as the 'experimental' or 'un-catalogued' 1958 Cape Dorset prints at the HBC's Winnipeg store was advertised as 'An important event at the Bay's Historical Exhibit, 4th floor – the first showing of a new Eskimo handicraft – stone engravings made in the Arctic by Eskimo carvers',[27] creating a sense of continuity between the craft designation of Inuit sculpture and the new 'Eskimo handicraft' of stone engraving. For members of the Department of Northern Affairs and National Resources, the success of Inuit printmaking could come only by firmly establishing it within a larger fine-art tradition and distinguishing it from both craft and the souvenir market. As a result, the Department set its sights on the art museum as the venue from which to launch Inuit printmaking and to closely control the supply of prints by establishing annual print releases. To this end, the early prints were removed from circulation – despite their commercial success – and the studio at Cape Dorset refocused its production model to more closely resemble a fine-art print studio.

The official launch of the first Cape Dorset print collection took place at the Montreal Museum of Fine Arts (MMFA) in February 1960.[28] At a lavish dinner attended by the Minister of Northern Affairs and National Resources alongside other government officials, the MMFA's Director Evan Turner praised the new direction in Inuit art. The choice of venue clearly underscored the government's belief that the marketing of this new form of Inuit cultural production must occur through the art museum and select commercial galleries in order to clearly establish it within the art market. To further achieve this end the Canadian Eskimo Art Committee (CEAC) was established in 1961 with a mandate to oversee the development of Inuit graphics production and to control its circulation.[29] The committee consisted of Turner as Chairman, Paul Arthur, Managing Editor of *Canadian Art*, Alan Jarvis, Editor of *Canadian Art* and former Director of the National Gallery, Budd Feheley, President of TDF Artists Limited in Toronto and Norman Hallendy from the Department of Northern Affairs. The Committee's tasks were to select prints sent for approval by the nascent art cooperatives in the Arctic, to authenticate all artwork produced in the Arctic, and to advise on appropriate marketing strategies for all graphic work produced in the Arctic, including the annual collections chosen for release. In this latter role, the Committee saw its task as closely aligned to standard practices among fine-art printmakers.

With the establishment of the CEAC and its powerful art-industry insiders, the shift in designation of Inuit cultural production from 'craft' to 'art' was complete. Even as it was being launched in the art gallery, however, Inuit printmaking was still being presented as a 'primitive' art form. In the catalogue accompanying the MMFA launch, James Houston emphasized the relationship between the traditional way of life of the Inuit and the images appearing in their prints, noting the remoteness and isolation of Arctic peoples and their resulting difference from the Southern consumers he was addressing. This reliance on the primitive character of the artists to market Inuit prints can also be seen in the concern voiced by Evan Turner to the Department of Northern Affairs in 1962 that the financial success of the print programme could 'endanger the convictions behind the source of that income, namely the actual works of art',[30] and that the government should develop a policy for the proper management of the funds coming into the printmaking communities.

Despite the presence of significant Canadian art-world figures on the CEAC, the success of the printmaking programme must be placed firmly in the hands of the people who enabled the sale of individual prints. As the reports by the women's committees of the Edmonton and Vancouver Art Galleries indicate, while Inuit graphic art was launched in the museum, it circulated primarily through the gift shops initiated and managed by these societies. In facilitating the entry of Inuit art into gallery gift shops, and thus making them available to a broad public, volunteer women's societies played an important transactional role in altering the object status of these works and thus fulfilling the aesthetic potential imagined for Inuit art by the federal government. While the commodity status of Inuit art remained a dominant feature of both the sculptures and the prints, the aura of the art museum gift shop provided the aesthetic framework that was absent from the HBC stores' Canadiana boutiques, and was arguably missing from the CHG shops. It is clear from the CEAC files that the government targeted art museum gift shops as the most appropriate venue to bring Inuit prints to market because of the exclusivity and aesthetic associations attributed to the art gallery gift shop; but it is also likely that Department officials were reassured that these shops were staffed by the same kind of women who, through their leadership of the CHG, had ensured the careful distribution of Inuit sculpture in the 1940s and 1950s.

Women's societies would be formative in yet another central transaction in the history of Inuit art: its move into the permanent collections of the country's art galleries. At the Edmonton Art Gallery, the agreement to sell the newly introduced Inuit prints in 1960 was followed in 1969 by a decision to spend $2,000 on the acquisition of Inuit prints and sculptures for the Gallery's permanent collection.[31] These would be the first works by Inuit artists to be acquired by the Edmonton Art Gallery, and the first works of art to be selected directly by the Women's Society for purchase (Figure 5.4). In Winnipeg, a few Inuit sculptures were purchased by the Winnipeg Art Gallery (WAG)'s Women's Committee in 1959 but in 1961 they used a significant proportion of the funds raised

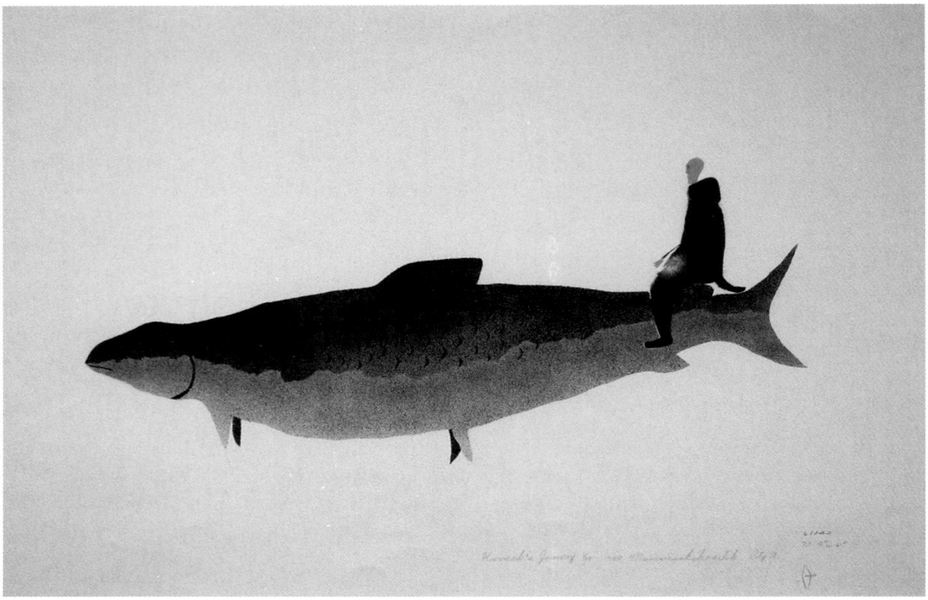

5.4 Victoria Mamnguksualuk, *Keeveok's Journey*, 1969, Stencil on paper.

that year to purchase 134 sculptures from George Swinton's collection: a decision that effectively jump-started the WAG's now substantial collection of Inuit art.[32] The determination of both women's committees to directly purchase Inuit art is important. While much of the money raised by art museum women's committees went to the purchase of art works for the permanent collection, the majority of societies had no input into decisions around acquisitions. A significant exception is the AGO's Women's Committee, which initiated the purchase of contemporary Canadian, American and British art for donation to the AGO's collection because of their impressive personal connections with art dealers in New York and elsewhere.[33] Other societies made occasional purchases from the yearly sales of contemporary art that they organized – the WAG's Women's Committee regularly bought works from its annual 'Winnipeg Shows' for donation to the WAG's permanent collection[34] – but for the most part all funds raised by these groups specifically for acquisitions were handed over to the institutions' directors and curators with no strings attached and surprisingly few complaints from the women.

Given this lack of collecting activity, it is interesting that the one area in which women's societies were willing – and, it should be noted, *permitted* – to assert their autonomy was the acquisition of Inuit art. It could be argued that many of the women involved had a degree of informal expertise in Inuit art since they had not only handled such objects as part of their work in the gift shops, but had made the decision early on to include Inuit art in their shops' inventories. The legacy of the women of the CHG cannot be overlooked here. The Guild's leadership in introducing Inuit carvings and crafts to the Canadian public was achieved through a network of women's organizations whose members staffed the Guild's shops across the country or provided opportunities for sales of carvings and prints through museum shops and craft sales. Anecdotally, it would also appear that many of the women collected Inuit art themselves, although perhaps not at the level or in the numbers achieved by such male collectors as George Swinton, Harry Winrob or Ian Lindsay, whose donations have been celebrated with major exhibitions and large catalogues.[35] I would argue, however, that women's role in ensuring the circulation of Inuit art during the formative years of its development, and in promoting it as a highly collectible art form, is key to the current position enjoyed by Inuit art in the contemporary art museum. As sales outlets, the Guild's shops and museum gift shops were predominantly sites of economic consumption, but their association with fine craft and fine art associated these acts of consumption with the aesthetic values of the art museum. As a result of this association, the museum gift shop was foundational in establishing the aesthetic legitimacy of Inuit art, functioning as a transformational space where the shift from pure commodity to art form could be realized. It was this process that undoubtedly underscored the decision by the Department of Northern Affairs and National Resources to focus its marketing of Inuit prints through art museums.

In her consideration of the role of female patronage in the development of Southwest American Indian art in the 1920s and 1930s, Molly H. Mullin has argued that women deployed dominant assumptions about their consumption habits and predilection for shopping to mobilize broader awareness of Indigenous arts and land rights. In other words, women's consumption of Native art objects through tourism, in their own writing, and especially through purchases, was instrumental in bringing the cultural production of this area to national attention as objects of fine art. More importantly, Mullin argues that women's exclusion from decision-making positions in established fine-arts organizations led many to explore marginalized and undervalued forms of aesthetic production such as Indigenous arts, and to advocate for their place within the institution.

> While middle- and upper-class women have been positioned as specialists in areas of taste and consumption, they have nonetheless had difficulty gaining access to positions of substantial authority in art fields with a more established institutional structure. Disadvantaged by their gender status on their home turf, these women managed to increase their cultural capital by entering and commandeering an area both literally and figuratively at the periphery of the political and economic system.[36]

As the creators and managers of art museum gift shops, volunteer women's societies were instrumental in creating both the spaces of consumption and the means of circulation through which Inuit art achieved a high degree of legitimacy in the art museum. The creation of community produced in and across these spaces of sociability, as well as the strength of the voluntary women's network, provided opportunities for many women to actively participate in an art world from which they were professionally excluded. While it is tempting to dismiss the gift shop as little more than the locus of commercialism, the federal government's decision to introduce the new medium of Inuit printmaking through museum gift shops speaks to the quality of these venues as well as to the recognition of the abilities and resources of the women who staffed them. That the women themselves translated this commercial venture into an opportunity to make their own mark on the collections of the art institution is a further testament to their desire to position themselves as active members of the art museum.

While women's societies were fundamental to the acquisition of Inuit art by Canadian museums, this contribution has been all but forgotten, as have the women themselves. What I hope I have successfully argued here is that we must acknowledge the work of volunteer women's societies in creating the commercial spaces that framed Inuit cultural production in predominantly aesthetic terms and in exercising strong initiative in ensuring the inclusion of these objects within art galleries' permanent display. Inuit art – now considered a central component of Canadian art history and a staple of art museum collections across the country – has a fascinating historiography; we need to attend to women's roles in bringing it from the gift shop to the permanent collection.

Sources Cited

Goetz, Helga, 'The Role of the Department of Indian and Northern Affairs in the Development of Inuit Art', unpublished manuscript, Ottawa: Inuit Art Section of the Department of Indian and Northern Affairs, 1985.

Gustavison, Susan, *Arctic Expressions: Inuit Art and the Canadian Eskimo Arts Council 1961–1989* (Kleinburg: McMichael Canadian Art Collection, 1994).

Igloliorte, Heather, 'Influence and Instruction: James Houston, Sunnyuksuk: Eskimo Handicrafts, and the Formative Years of Contemporary Inuit Art', MA Thesis (Carleton University, Ottawa, 2006).

Laverty, Paula, *Silk Stocking Mats: Hooked Rugs of the Grenfell Mission* (Montreal and Kingston: McGill-Queen's University Press, 2005).

Logan, Kathleen, 'Acquiring Authority at the Art Gallery of Ontario and the Agnes Etherington Art Centre: Canadian Women and the Cultural Volunteer Experience', MA Thesis (Ottawa: Carleton University, 1997).

Macleod, Dianne Sachko, *Enchanted Lives, Enchanted Objects: American Women Collectors and the Making of Culture 1800–1940* (Berkeley, CA: University of California Press, 2008).

McCarthy, Kathleen D., *Women's Culture: American Philanthropy and Art, 1830–1930* (Chicago, IL: University of Chicago Press, 1991).

McLeod, Ellen Easton, *In Good Hands: The Women of the Handicrafts Guild* (Montreal and Kingston: McGill-Queen's University Press, 1999).

McMaster, Gerald (ed.), *Inuit Modern: The Samuel and Esther Sarick Collection* (Toronto: Art Gallery of Ontario and Vancouver: Douglas & McIntyre, 2011).

Mullin, Molly H., *Culture in the Marketplace: Gender, Art, and Value in the American Southwest* (Durham, NC and London: Duke University Press, 2001).

Nosan, Gregory, 'Women in the Galleries: Prestige, Education, and Volunteerism at Mid-Century', *Art Institute of Chicago Museum Studies*, 29/1 (2003), pp. 46–71, 92–5.

Potter, Kristin K., 'James Houston, Armchair Tourism, and the Marketing of Inuit Art', in W. Jackson Rushing (ed.), *Native American Art in the Twentieth Century: Makers, Meanings, Histories* (London and New York: Routledge, 1999), pp. 39–56.

Rushing, W. Jackson, 'Marketing the Affinity of the Primitive and the Modern: René d'Harnoncourt and *Indian Art of the United States*', in Janet Berlo (ed.), *The Early Years of Native American Art History* (Seattle, WA: University of Washington Press and Vancouver: University of British Columbia Press, 1992), pp. 191–236.

Russell, Bruce Hugh, 'Aboriginal Art', in Jacques Des Rochers (ed.), *Quebec and Canadian Art: The Montreal Museum of Fine Arts' Collection*, vol. 1 (Montreal: Montreal Museum of Fine Arts, 2011), pp. 80–85.

Scoles, Diane, 'The Winnipeg Show 1955–1970: Local and National Significance', MA Thesis (University of Victoria, Victoria, 1995).

Staples, Annalisa R., Ruth Ethel McConnell and Patricia Alice McCormack, *Soapstone and Seed Beads: Arts and Crafts at the Charles Camsell Hospital, a Tuberculosis Sanitorium* (Edmonton: Provincial Museum of Alberta, 1993).

Vorano, Norman, 'Introduction: Early Printmaking in Cape Dorset', *Inuit Prints: Japanese Inspiration: Early Printmaking in the Canadian Arctic* (Gatineau: Canadian Museum of Civilization, 2011), pp. 3–12.

Vorano, Norman, 'Inuit Art in the *Qallunaat* World: Modernism, Museums and the Popular Imaginary, 1949–1962', PhD Dissertation (University of Rochester, Rochester, 2007).

Wasson, Haidee, 'Every Home an Art Museum: Mediating and Merchandising the Metropolitan', in Charles Acland (ed.), *Residual Media* (Minneapolis, MN: University of Minnesota Press, 2007), pp. 159–84.

Watt, Virginia, 'The Beginning', *The Permanent Collection: Inuit Arts and Crafts c1900–1980* (Montreal: Canadian Guild of Crafts, Quebec, 1980), pp. 11–14.

Whitelaw, Anne, 'Professional/Volunteer: Women at the Edmonton Art Gallery 1923–1970', in Kristina Huneault and Janice Anderson (eds), *Rethinking Professionalism: Essays on Women and Art in Canada, 1850–1970* (Montreal and Kingston: McGill-Queen's University Press, 2012), pp. 357–79.

Whitelaw, Anne, 'Women, Museums and the Problem of Biography', in Kate Hill (ed.), *Museums and Biographies: Stories, Objects, Identities* (Woodbridge: Boydell and Brewer, 2012), pp. 75–86.

Wight, Darlene Coward, 'The Handicrafts Experiment, 1949–53', *The First Passionate Collector: The Ian Lindsay Collection of Inuit Art* (Winnipeg: Winnipeg Art Gallery, 1991), pp. 45–92.

The Swinton Collection of Inuit Art (Winnipeg: Winnipeg Art Gallery, 1987).

The Harry Winrob Collection of Inuit Sculpture (Winnipeg: Winnipeg Art Gallery, 2008).

Notes

1 'Inuit' is the proper term to refer to the Indigenous peoples of the Arctic. The term 'Eskimo' was in common usage by Southern peoples throughout the nineteenth and much of the twentieth

centuries. In this essay, I use the term 'Eskimo art' with quotation marks to highlight how cultural production by Inuit was known and referred to at the time. In all other instances I use the more culturally appropriate term 'Inuit art'.

2 The history of the early years of Inuit art has been told in many publications. Information for the following section was drawn from Helga Goetz, 'The Role of the Department of Indian and Northern Affairs in the Development of Inuit Art', unpublished manuscript (Ottawa, 1985); Heather Igloliorte, 'Influence and Instruction: James Houston, *Sunnyuksuk: Eskimo Handicrafts*, and the Formative Years of Contemporary Inuit Art', MA Thesis (Carleton University, 2006); Gerald McMaster (ed.), *Inuit Modern: The Asmuel and Esther Sarick Collection* (Toronto, 2010); Norman Vorano, 'Inuit Art in the *Qallunaat* World: Modernism, Museums and the Popular Imaginary, 1949–1962', PhD Dissertation (University of Rochester, 2007); Darlene Coward Wight, 'The Handicrafts Experiment, 1949–53', in *The First Passionate Collector: The Ian Lindsay Collection of Inuit Art* (Winnipeg, 1991), pp. 45–92.

3 Letter from R.G. Robertson, Deputy Minister Department of Northern Affairs and National Resources. to C.J.G. Molson, Chairman. Canadian Handicrafts Guild Shop Committee, 4 February 1955. CHG Archives C10 D1 081 1955.

4 Canadian Handicrafts Guild Annual Meeting, 1939, cited in Virginia Watt, 'The Beginning', in *The Permanent Collection: Inuit Arts and Crafts c1900–1980* (Montreal, 1980), p. 11.

5 The Guild leadership recognized early on that placing men in positions of authority was key to building a sense of legitimacy around its activities. As a result, several men – including C.J.G. Molson, Chairman of the Shop Committee in the 1950s – held high positions in the Guild. See Ellen Easton McLeod, *In Good Hands: The Women of the Handicrafts Guild* (Montreal and Kingston, 1999), pp. 172–3.

6 Wight, 'The Handicrafts Experiment', p. 50.

7 Goetz, 'The Role of the Department of Indian and Northern Affairs', p. 47.

8 Letter from Mrs H.A. Cave, Chairman, Sale Committee of Women's Committee of the Windsor Art Association, to Martha Peters, Canadian Handicrafts Guild, 12 October 1954. CHG Archives C10 D1 073 1954.

9 Letter from C.J.G. Molson, Canadian Handicrafts Guild to R.C. Ingram, Executive Assistant. Northern Stores Department, Hudson's Bay Company, 9 July 1960. CHG Archives C10 D1 117 1960. It should be noted that Gander airport was an important refueling stop for many transatlantic flights and was therefore a prime location for the sale of Canadiana (souvenirs), a designation that included Inuit carvings.

10 Letter from C.J.G. Molson to S.G. Hazell, Executive Vice President, Canadian Handicrafts Guild, Ontario Branch, 4 January 1957. CHG Archives C10 D1 088 1955.

11 Kristin K. Potter, 'James Houston, Armchair Tourism, and the Marketing of Inuit Art', in W. Jackson Rushing (ed.), *Native American Art in the Twentieth Century: Makers, Meanings, Histories* (London and New York, 1999), pp. 39–56.

12 See chart appended to a letter from C.J.G. Molson to C.A. Stewart of the HBC 17 April 1965. Based on statistics of carvings sold at the Guild's Montreal shop in August and December 1964, the chart demonstrated the greater popularity of the smaller, less expensive carvings. Molson argued that the basis of production – what to encourage the carvers to produce – should be 70 per cent in the \$1–\$8 range, and 10 per cent in each of the \$8–\$15, \$15–\$25 and \$25–\$50 ranges, with nothing over \$50 except 'really good pieces'. This chart came at the end of a long discussion about the scale of the sculptures, and strong suggestions from both the Guild and the HBC that carvers should be actively discouraged from producing large (and thus more expensive) pieces, as there was virtually no market for them. CHG Archives C10 D2 171 1965.

13 Letter from Carl Fox, manager of The Gallery Shop (Brooklyn Museum) to the Canadian Handicrafts Guild Shop, 20 September 1954. CHG Archives C10 D1 073 1954. Interestingly, Fox would become the manager of the gift shop at the Smithsonian in 1967 and would contact the Canadian Eskimo Arts Council to enquire about the sale of Inuit prints.

14 See W. Jackson Rushing, 'Marketing the Affinity of the Primitive and the Modern: René d'Harnoncourt and *Indian Art of the United States*', in Janet Berlo (ed.), *The Early Years of Native American Art History* (Seattle and Vancouver, 1992), pp. 191–236.

15 Haidee Wasson, 'Every Home an Art Museum: Mediating and Merchandising the Metropolitan', in Charles Acland (ed.), *Residual Media* (Minneapolis, 2007), pp. 159–84.

16 I have published on aspects of the work of volunteer women's societies in museums in 'Professional/ Volunteer: Women at the Edmonton Art Gallery 1923–1970', in Kristina Huneault and Janice Anderson (eds), *Rethinking Professionalism: Essays on Women and Art in Canada, 1850–1970* (Montreal and Kingston, 2012), pp. 357–79; and in 'Women, Museums and the Problem of Biography', in Kate Hill (ed.), *Museums and Biographies: Stories, Objects, Identities* (Woodbridge, 2012), pp. 75–86. See also Gregory Nosan, 'Women in the Galleries: Prestige, Education, and Volunteerism at Mid-Century', *Art Institute of Chicago Museum Studies*, 29/1 (2003), pp. 46–71 and 92–5.

17 Projects Committee Report Re Shop, attachment to the Minutes of the Women's Committee of the Art Gallery of Toronto, 26 October 1953, Art Gallery of Ontario, EP Taylor Library and Archives.

18 Ibid.

19 Women's Society of the Edmonton Museum of Arts, *Annual Report 1944–45*, Edmonton Art Gallery fonds, box 14, vol. 3, City of Edmonton Archives.

20 The Grenfell Mission was founded in 1893 to assist impoverished fishers in Newfoundland and Labrador. Wilfrid Grenfell recruited Jesse Luther in 1905 to encourage women of the region to produce traditional handicrafts for sale. The Camsell Hospital in Edmonton was one of several hospitals that took in Inuit suffering from tuberculosis. In 1946 crafts were introduced as part of the hospital's Occupational Therapy programme, and Inuit patients were encouraged to produce objects based on traditional designs. Only some patients were allowed to carve, however, due to the dust produced by manipulating soapstone. See Paula Laverty, *Silk Stocking Mats: Hooked Rugs of the Grenfell Mission* (Montreal and Kingston, 2005) and Annalisa R. Staples, Ruth Ethel McConnell and Patricia Alice McCormack, *Soapstone and Seed Beads: Arts and Crafts at the Charles Camsell Hospital, a Tuberculosis Sanitorium* (Edmonton, 1993).

21 It should be underlined that only the crafts obtained from the Camsell Hospital were produced by Inuit. The objects sent out from the Grenfell Mission employed designs that used Arctic or 'Eskimo' motifs but were not based on traditional designs or made by Inuit producers. I would like to thank Heather Igloliorte for pointing out this important detail about the Grenfell Mission crafts.

22 For a thorough analysis of the introduction of printmaking in the Arctic see Norman Vorano, 'Introduction: Early Printmaking in Cape Dorset', in *Inuit Prints: Japanese Inspiration: Early Printmaking in the Canadian Arctic* (Gatineau, 2011), pp. 3–12.

23 Goetz, 'The Role of the Department of Indian and Northern Affairs', p. 51.

24 Ibid., p. 56.

25 Art Rental and Sales was a venture initiated by Junior League groups in the mid-1950s in many Canadian cities in conjunction with the volunteer women's committees of the art gallery. Paintings, prints and occasionally sculptures would be available for rental by members of the

public for periods of up to one year, renewable. Renters would pay a monthly fee, with the cost of the rental deducted from the price of the work should a purchase eventually be made. Designed to increase public exposure to fine art, Art Rental programmes were very popular across the country both to the public and to the largely local artists who benefited from getting their work seen. Art Rental and Sales is now a major service offered in virtually all Canadian museums.

26 Emphasis mine. Minutes of the Women's Society of the Edmonton Art Gallery, 14 March 1963. Edmonton Art Gallery fonds, box 12, vol. 2, City of Edmonton Archives.

27 The Winnipeg *Free Press*, 13 December 1958. See also Vorano, 'Introduction', p. 7.

28 The MMFA was an entirely appropriate venue from which to launch the Inuit graphics programme as that institution had a history of collecting work by First Nations artists dating back to the 1920s when F. Cleveland Morgan was both President of the Art Association of Montreal's Board of Trustees and an active member of the Canadian Handicrafts Guild. See Bruce Hugh Russell, 'Aboriginal Art', in Jacques Des Rochers (ed.), *Quebec and Canadian Art: The Montreal Museum of Fine Arts' Collection*, vol. 1 (Montreal, 2011), pp. 80–85.

29 The Canadian Eskimo Arts Committee was renamed the Canadian Eskimo Arts Council in 1964. For a thorough examination of the work of the CEAC, see Susan Gustavison, *Arctic Expressions: Inuit Art and the Canadian Eskimo Arts Council 1961–1989* (Kleinburg, 1994).

30 Evan Turner, Memorandum to other members of the [CEA] Committee, 16 April 1962, Appendix A, Minutes of Meeting of the Canadian Eskimo Arts Committee, 6 May 1962. Canadian Eskimo Arts Council fonds, RG 85, vol. 2172, Library and Archives Canada. Turner continued with the statement that 'it would be generally agreed that the Eskimos themselves are not sufficiently sophisticated to make maturely such a decision [on their finances]'. In fact, the Department of Northern Affairs did not agree, and it conveyed to Turner and the other members of the Committee that the funds generated by the printmaking programmes were necessary for the economic health of the communities as a whole. Letter from R.H. Phillips to Turner, 1 May 1962 in ibid.

31 Minutes of the Women's Society of the Edmonton Art Gallery, 7 May 1969. See also Minutes of meetings on 11 June 1969, 10 September 1969 and 13 January 1971. Edmonton Art Gallery fonds, box 12, vol. 2, City of Edmonton Archives.

32 Cited in the finding aid to the Winnipeg Art Gallery fonds, Voluntary Committee sous-fonds. Winnipeg Art Gallery.

33 See Kathleen Logan, 'Acquiring Authority at the Art Gallery of Ontario and the Agnes Etherington Art Centre: Canadian Women and the Cultural Volunteer Experience', MA Thesis (Carleton University, 1997).

34 The Winnipeg Shows were annual exhibitions and sales of contemporary Canadian art organized by the Women's Committee of the Winnipeg Art Gallery between 1955 and 1970. In the 1950s and 1960s, these annual sales (also organized by the women's committees of the Art Gallery of Ontario, the Vancouver Art Gallery and the Montreal Museum of Fine Arts) constituted one of the primary sites for the circulation and consumption of contemporary Canadian art and were highly valued by many artists. See Diane Scoles, 'The Winnipeg Show 1955–1970: Local and National Significance', MA Thesis (University of Victoria, 1995).

35 See *The Swinton Collection of Inuit Art* (Winnipeg, 1987); *The Harry Winrob Collection of Inuit Sculpture* (Winnipeg, 2008); *The First Passionate Collector: The Ian Lindsay Collection of Inuit Art.*

36 Molly H. Mullin, *Culture in the Marketplace: Gender, Art, and Value in the American Southwest* (Durham, NC and London, 2001), pp. 88–9. Kathleen D. McCarthy and Dianne Sachko Macleod have also demonstrated women's roles as collectors of modern art at the turn of the twentieth century in legitimating that form of art in the art museum. See McCarthy, *Women's*

Culture: American Philanthropy and Art, 1830–1930 (Chicago, 1991) and Macleod, *Enchanted Lives, Enchanted Objects: American Women Collectors and the Making of Culture 1800–1940* (Berkeley, CA, 2008).

6

Benevolence, Revival and 'Fair Trade': An Historical Perspective

Janice Helland

During the nineteenth century, philanthropy or 'active benevolence' was considered a vocation suited to women who wanted to be engaged in meaningful activities but who could not be seen to 'work for profit'; thus 'it was deemed the leisured woman's most obvious outlet for self-expression'.[1] However, this 'tradition of philanthropy' also led many women, as Maria Luddy contends, 'to an understanding of the economic and social causes of poverty'.[2] In this chapter I suggest that this understanding led some late nineteenth-century activities, usually referred to as philanthropic, to become historical examples of 'fair trade' projects although the term was not used at the time. According to Gavin Fridell, the term 'fair trade' did not arise until after World War II when it became associated with a 'significant development project' that sought 'to address underdevelopment in the South by challenging the unequal terms of exchange for Southern handicrafts and commodities'.[3] Fridell proposes that the historical movement originated in the 'the inter-war period, 1918 to 1939';[4] Geoff Moore, on the other hand, suggests that the movement might be able to trace its beginnings to the 'co-operative movement in the late nineteenth century'.[5]

Historical Fair Trade

Moore accepts FINE's definition of fair trade: 'Fair trade organizations (backed by consumers) are engaged in actively supporting producers, awareness raising and in campaigning for changes in the rules and practices of conventional trade'.[6] And, following that definition, I argue that certain historical Irish and Scottish organizations that avoided middlemen and returned profits to artisans can be discussed as fair trade projects. The women who supported, promoted and organized these ventures can be characterized as philanthropic in Prochaska's sense of the word; but they also proposed a market that sought 'to establish a bond between producers and the buyers of their goods' and thereby they disrupted the 'impersonal nature of global markets' and substituted 'values of community and solidarity'.[7] Concerns about artisan communities, middlemen, returning profits to makers and finding markets for products were voiced by, for example, Londoner Alice Hart when she visited Donegal in the 1880s, Ishbel, Countess of Aberdeen, when she organized the Irish Industries Association in 1886 and Millicent, Duchess of Sutherland, when she championed Highland crafts during the 1890s.

Following Lynne Milgram's and Kimberly Grimes's assertion that 'factory work seldom provides the independence or security many young women seek' and that 'local craft production can lead to more self-reliance and independence for women',[8] this chapter will examine one nineteenth-century organization, the Donegal Industrial Fund, as a fair trade network while accepting the problems with this approach. It is important to note that the emphasis here is on a fair trade network and not on the fair trade movement, which includes 'a commitment to a vision of an alternative international trading system with strong international market regulation'.[9] An emphasis on fair trade discourse, even that of a network, can obscure 'structural inequality' and instead support a 'liberal, depoliticised vision of cultural difference' wherein 'ethnic branding, tradition, and authenticity' become the key features in the marketing of the crafts.[10] So why frame a discussion of a late nineteenth-century organization in this way? Might such an approach result in a new way of thinking about philanthropic historical projects organized by women? Does such a shift in focus enable a greater discussion of process and allow more scope for discussion of community and agency? Might a reconfigured explication of the Donegal Industrial Fund position the organization within a context of global consumption and, in that context, enable a discussion of the internal and external consumption of Irish-made craft as participating in interrelated exchanges and interactions? This approach to the consumption of Irish craft borrows from Joanna de Groot's analysis: she 'repositions consumption as an *active* element in the complex processes of material change, interacting with labour, skill and entrepreneurship, physical and financial resources, and forms of organisation of investment and production'. Thus she considers 'social relations, cultural forms and political institutions' as activities that shape consumption and, in doing so, contribute to the 'interaction of material, political and cultural aspects of past human experience'.[11]

This chapter looks back to the 1880s when energetic arts and crafts enthusiast, Londoner Alice Hart (1848–1931), established a non-profit organization to promote and sell objects, particularly textiles, crafted in Donegal. She initiated her project, the Donegal Industrial Fund, after she had travelled through the poverty-stricken countryside in north-western Ireland and subsequently offered artisans an opportunity to improve the quality of their goods, promised to transport the products to viable markets, and to return profits to the makers.

The Donegal Industrial Fund was one of a number of so-called philanthropic organizations established in Britain and Ireland during the latter part of the nineteenth century. Hart's objective was to establish what today would be called a 'fair trade network' that would benefit artisans who were otherwise at the mercy of middlemen. Thus the Fund will be considered an early example of a 'fair trade' network or an alternative trade organization (ATO) that focused on the revival or retention of traditional craft while, at the same time, it returned profits to Irish communities and sought global markets. Coincidentally, a report on the 'standard of industrial design in Ireland' in the 1960s published by an international team of consultants corroborates Moore's suggestion about the historical roots of fair trade and retains Hart's interest in the revival of 'tradition': 'the better designed products were based on traditional arts and crafts interpreting the Irish tradition'. The report went on to describe Donegal tweeds 'as the most valuable facet of Ireland's textile industry'. In his discussion of arts and crafts in Donegal, Seán Beattie contents that the glowing report was 'an enduring tribute to the standards pioneered throughout the previous century'.[12]

Revival of artisanship certainly figured in discussions about these historical craft organizations, as did philanthropy, benevolence and the relationship between craft making and political stability. Thus, while I shall reconfigure one organization, the Donegal Industrial Fund, within the concept of a historical 'fair trade' network, I shall, at the same time, recognize that beneficence (if not philanthropy) played a role in the functioning of the Fund. One might, however, query the view that the Fund was a conservative organization that sought to preserve the morality of rural life and instead discover agency in the decision to commit to the production of craft. Additionally, while Alice Hart's social status as the daughter of a wealthy mercantile family and the wife of a successful physician placed her hierarchically above the Donegal craft makers, the slight anecdotal information available suggests that she was respected and her efforts were perceived as beneficial to the community.[13]

Donegal Industrial Fund

With these thoughts about fair trade networks and consumption in mind, I now turn to the Donegal Industrial Fund. Alice Hart reiterated the story of the founding of the Fund in her speeches (some of which were published as pamphlets), in articles she wrote for the press, in interviews and in numerous magazine and newspaper articles published in Great Britain, Ireland and the United States. The narrative began with an early trip to Ireland in 1879 when she was 'bent solely on pleasure and holiday making' but, at the same time, she began to ask questions about the Irish people, their land and their industry. Her interest, however, did not become a project to assist craftspeople in Ireland until 1883, when she and her husband, Ernest Hart, 'went on a tour of inquiry into Donegal and county Mayo'; their purpose was to 'ascertain the actual condition of, and the possible means of helping, the peasantry in the congested districts, the scenes of recurring famine'.[14] By December of that year, 'a fund was set on foot' by Alice Hart, 'with the object of reviving and promoting the cottage industries of Ireland'. It was to focus upon the 'manufacture of homespun and handworked woollen and linen fabrics' and quickly expanded to include 'spinning, weaving, embroidering, and dyeing'.[15] Within two years, in 1885, Hart celebrated the opening of a London storefront, Donegal House, near Oxford Street, with an exhibition of homespuns, tweeds, embroideries and woollen fabrics.[16] As Alice Hart promoted and advanced her project, she quickly exceeded the boundaries of national trade and prepared to send Irish objects to international venues, particularly to the United States, thereby entering the global market in her effort to secure consumers for Irish craft.

The Fund was first introduced to American consumers when Hart opened an 'exhibition and sale of laces, fabrics and embroideries' in New York in 1886, 'courtesy of the Associated Artists', the fledgling organization that started operating in 1883 under the direction of Candace Wheeler. The venue was perfect for Hart's display; the association itself held views very close to Hart's own, which included designing textiles and 'encouraging women in their quest to become self-supporting professional designers and artists'.[17] The *New York Times*, in its announcement about the exhibition and sale, wrote that 'Irish peasant girls and destitute ladies' could produce the work during the winter months while the Fund secured a market. Additionally, the newspaper lauded the business side of the venture: the 'enterprise has been conducted on wise business principles and has succeeded beyond expectations'. According to the *Times*, the invoice of goods displayed at the rooms of the Associated Artists was 'worth nearly $3,000';[18] after expenses, the profits were returned to the craft makers. Within a few years, the Fund expanded its American markets to include Philadelphia,[19] Chicago and San Francisco.

The objects prepared for New York were also admired in London. Queen Victoria ordered curtains 'made and embroidered by the workers of the fund for Windsor Castle similar to those' she had seen 'which had been specially hand-woven, designed and embroidered for the Associated Artists of New York'.[20] Royal and aristocratic patronage secured markets, ensured press coverage and guaranteed interest in both product and artisan (no different really than press coverage for Sarah Burton and London's Royal School of Needlework after the recent Royal Wedding).[21]

Soon after the New York success, in the spring of 1887, Countess Spencer (vicereine of Ireland 1882–85) hosted an exhibition of Irish craft produced under the auspices of the Fund in her London townhouse. The display attracted a number of British and Irish aristocrats, including the Princess of Wales, the Duchess of Marlborough and the Countess of Carnarvon. In its comments on the exhibition, the *Irish Times* claimed that the Fund gave 'skilled employment' to Irish cottagers and thus demonstrated 'that the starving peasantry of the far West, in the wild outlying regions of rugged Donegal, possess instinctive skill and artistic taste that qualify them to produce the textile fabrics for which Old Ireland was famous'. These textiles promoted and consumed in the metropole commanded 'a ready market in the most fashionable quarters'.[22]

To return to FINE's suggestion that one goal of today's fair trade movement is to 'establish a bond between producers and the buyers of their goods, rupturing the impersonal nature of global markets and substituting values of community',[23] we might also consider that this was a goal of the nineteenth-century Donegal Industrial Fund (DIF). Like today's fair trade, the DIF also devoted much of the discussion of its own practices to a dissemination of information about the districts in which the goods were produced. Hart focused her activities on villages or small rural communities in remote parts of Donegal where 'technical classes' could be offered. She looked to similar ventures that she had encountered when she travelled in Europe, particularly in the 'mountain districts of Bohemia' and in Prague. In Prague she met with a committee that had been in existence for 15 years and encouraged 'the village and hand-industries of the Erzgebirge and Riesengebirge'. She saw villagers who worked communally to produce specific crafts; for example, one village specialized in making musical instruments, another in lacemaking and, she told her audience, a similar scheme could be implemented in Donegal.[24]

Later, when she addressed an audience in Chicago in 1892, she characterized the people of Donegal as 'hardy, penurious, and industrious'. 'Their precarious methods of gaining a livelihood, and the necessity to depend upon themselves', said Hart, 'have combined to make them thrifty, long-headed, and independent'. Thus she conveyed a verbal image of a community of dedicated workers as she 'displayed to her audience many beautiful specimens of the work done by these peasants in silk, cotton, and flax embroidery, silk and linen lace of a wonderful texture and design, and specimens of homespun that rivalled the products of the best factories in the world'.[25]

Philanthropy, Beneficence or Fair Trade?

Although Hart generally was considered to have started the project as a philanthropic venture, one voice challenged that view by recognizing the soundness of the financial aspect and the 'self-dependence' accorded the craftspeople. Florence Fenwick Miller, a well-known public speaker, supporter of women's rights and columnist for the *Illustrated London News*, observed in her remarks about the 1887 exhibition that the Fund was 'saved from being a mere charity, dependent for its existence on sympathy with Irish poverty, by the originality of the work which it offers, and a sufficient sale which will make it self supporting'.[26] In Hart's lifetime, however, and Florence Fenwick Miller aside, the venture received criticism as philanthropy: Hart, according to an anonymous journalist who commented on the Fund's contribution to the 1885 Inventions Exhibition (London), might 'claim an honourable place among the inventions if she and her philanthropic friends could show an invention for making the helpless Irish peasantry help themselves … Nothing could be more depressing than to read the record of the industries set up and fostered by English capital and philanthropy', and, insisted the commentator, 'as long as these energetic ladies will push the go-cart themselves the cotters will probably consent to go in it'.[27] Comments such as this did little more than patronize the 'ladies' and demean the 'cotters' without providing an analysis of the exchanges and relationships. Additionally, the label of philanthropy does not correctly describe a project that engaged with the commercial marketplace. Jeffrey Brison, for example, in his study of American philanthropy as it related to Canada, discusses one-directional funding that flows from the benefactor or philanthropist to an organization or an individual, such as Andrew Carnegie's monetary support for the establishment of libraries in Canada.[28] Thus philanthropy is more accurately considered an aspect of giving devoid of economic exchange. If nineteenth-century so-called philanthropic organizations are closely examined, they can be found to fall into two categories: first, one-directional funding or assistance; and, second, market-oriented, non-profit funding based on an exchange of goods for money.

The concept of benevolence might, however, be considered an aspect of these same nineteenth-century projects, including the DIF; and benevolence was and remains a suspect endeavour. Helen Gilbert and Chris Tiffin reviewed the contradictions and complexities of benevolence in their introduction to *Burden or Benefit? Imperial Benevolence and Its Legacy* (2008), and their observations aptly translate to the DIF. Their discussion also offers insights into how and why benevolence has become historically suspect:

> virtually all forms of personal benevolence, even the most apparently altruistic, involve a structural relationship that situates the donor as a dominant, self-approving figure. Benevolence, then, is never simple, and its complications multiply exponentially when the case is not that of the individual within a contained culture, but rather that of an organization or nation acting across cultures.[29]

Alice Hart did reach across cultures with the expectation of communicating certain 'beliefs and practices' that were valued by her compatriots but not necessarily valued by producers of the objects exhibited for sale in the metropole. The participation of aristocratic 'ladies' also contributed to the gendering of the philanthropic venture, which, in turn, assured its place on the margins of the Arts and Crafts movement of which it was a part. An essay in a recent book that accompanied a major arts and crafts exhibition referred to cottage craft organizations such as the DIF as 'mainly rural, [a] world of craft as philanthropy, a world of village halls and ladies of the manor';[30] another author called the organizations the 'philanthropic arm'[31] of the Arts and Crafts movement.

However, if one considers craft such as that produced by artisans for the DIF, which, in turn, exhibited the objects widely and in numerous venues, then contemplation of today's fair trade networks does offer a way of disrupting the conventional story of some late nineteenth-century home arts – that is, not as a 'victory against neoliberal globalization' but rather as a 'voluntarist, nonstatist development strategy' that connected 'peasants, workers, and craftspeople' in an increasingly industrializing economy with those who could provide a market for hand-crafted products and contribute to a facilitation of artisanship. An examination of the exhibited objects of the DIF using the model of a fair trade network opens up alternative ways of discussing the objects, the spaces of display and the viewing and consuming audiences. The movement was as much about professional training and the development of skills as it was about a so-called anti-modern revival of peasant or lost arts. And a discussion of the DIF as a fair trade network enables a more complex analysis of what remained a colonial project but, at the same time, encouraged independent artisans. Additionally, the troubled narrative of gender and philanthropy can be disrupted by complicating the relationship between the DIF and its lower-class subjects who, for the most part, remain silent.

Although there is much evidence to indicate that English imperialism preferred self-supporting, working peasants to restless, politicized agitators, the DIF also offered artisans the opportunity to market their craft in large urban centres. The colonialism of the project was blatantly obvious, but the method of promotion and sale did prefigure fair trade. Accordingly, a discussion of the 'honourable livelihood' promised to distressed Irish needlewomen could be reconceptualized as a fair trade network, thereby complicating the space and disrupting the perception of nineteenth-century philanthropy. This may also enable historians to discard a simplistic narrative of oppression or even of resistance for a more nuanced approach that recognizes that cross-cultural encounters – and the Irish–English encounter is most definitely cross-cultural – are never neutral but 'assertive, contested, and dialogic'.[32] The objects produced in the geographical space of Donegal speak in this never-neutral space to the artisans, to advocates or facilitators such as Alice Hart, and to consumers.

The most frequently discussed objects written about in the nineteenth-century press were embroideries such as 'Dragon and Serpent' (1886) (Figure 6.1), a detail from the curtains designed for the Associated Artists of New York, first exhibited at the Edinburgh International Exhibition (1886), and then subsequently designed for Windsor Castle. Embroidery exhibited and sold by the Fund was done on hand-spun and hand-woven wool or linen; the thread was either silk or more frequently dyed and polished flax thread, and both ground and thread were hand-dyed.[33] In an 1887 speech at the Society of Arts, Alice Hart labelled the design on the curtains 'Kells embroideries' in which one could see 'the interlacing of the quaint Celtic devices, worked in silk on woollen cloth'. 'The whole of these curtains', declared Hart, 'are made by hand in Ireland, carded, spun, woven and embroidered by Irish hands; the design alone is by an English hand, Aimée Carpenter, of Croydon, to whose happy genius for design our embroidery work owes much of its artistic success; but the inspiration is here drawn from an Irish source'.[34] Specifically, many of the designs were inspired by the illuminated manuscript, the Book of Kells, which dates to approximately 800 BCE, which Hart viewed in Trinity College in Dublin.

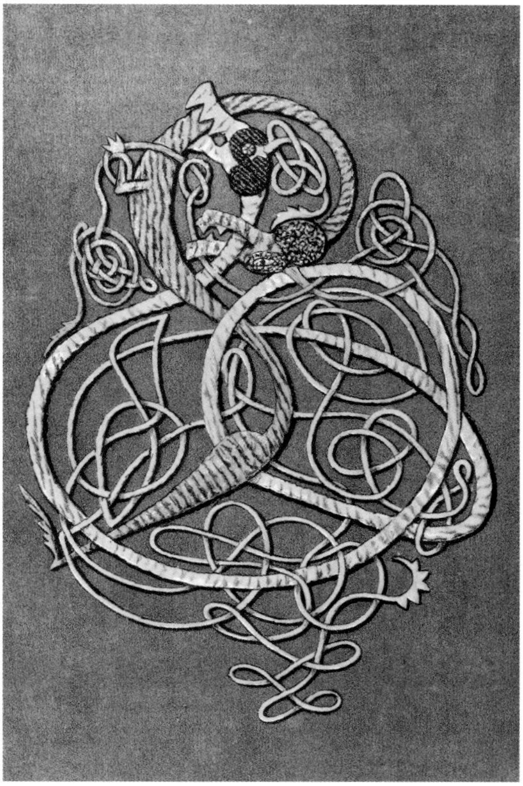

6.1 Kells Embroidery Dragon and Serpent, detail from the curtains for the Associated Artists of New York.

The dyeing of fabrics and threads included the use of local plants from Donegal; heather alone, for example, could produce an 'extensive range of colours,

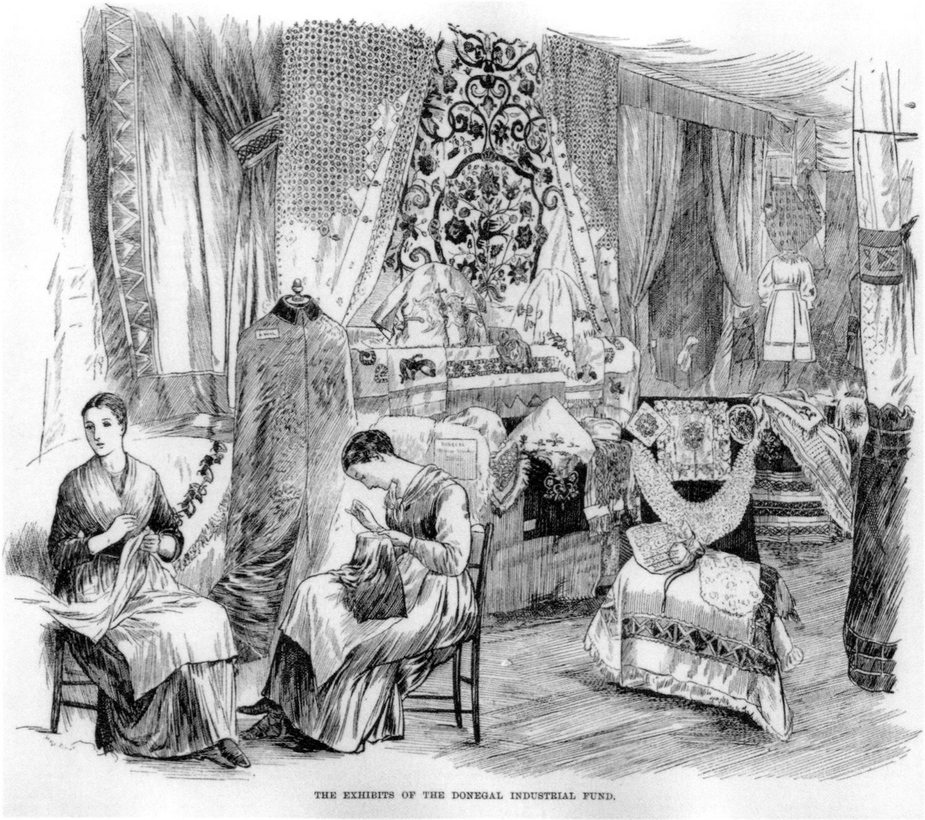

6.2 Kells Embroiderers' Cottage. The Irish Exhibition, Olympia, London, 1888. *Queen*, 28 July 1888, p. 108.

running through the gamut of pale buffs, yellows and browns', and with the added process of double-dyeing, the dyers could obtain 'vivid and delicate greens'.[35] As Hart reported in 1887, the dyes were made from 'the lichens and mosses of their own bogs'. Hart even audaciously promoted the Fund's work at the 1887 court Drawing Room when she wore a dress and train of hand-woven linen embroidered by Irish women with the Donegal trademark style 'Kells'. The conservative women's magazine, *The Lady*, commented: 'Whether this dress is very beautiful is a matter open to question; there can be no question as to uniqueness, nor as to the courage and originality displayed by Mrs Hart in wearing it'.[36]

During the 1888 Irish Exhibition in London a room was devoted to the embroideries: the Kells Embroiderers' Cottage (Figure 6.2) displayed linen and flax threads, dyed woollens and silk; the textiles were covered with dragons, serpents, floral and geometric designs in what one journalist dubbed 'a rainbow iridescence'.[37] To return to the language of fair trade, the materiality of the embroidered object thus spoke of 'ethnic branding, tradition and authenticity'.[38] Is then fair trade discourse just as dismissive of the DIF as philanthropic discourse? I would argue 'no' because it opens up a new avenue of discussion that, in turn, encourages a more complete and more complicated analysis of Hart's project. Additionally, the community of artisans can be explored in connection with both London's Irish Exhibition (1888) and the Chicago World's Fair (1893). Hart participated as advocate, promoter and organizer in both venues.

The 'Fair' Communities

The DIF had proved successful when it exhibited at London's Health Exhibition of 1884 – it received two medals for 'innocuous dyeing and hygienic woollen clothing' – and it obtained a silver medal at the New Orleans International Exhibition (1885),[39] but Hart's participation in London's Irish Exhibition of 1888 moved far beyond the inclusion of displays to the construction of a simulated Irish village complete with cottages and workers (Figure 6.3).[40]

The exhibition was one in a series of 'single-subject international exhibitions'[41] such as the Colonial and Indian Exhibition (1886), 'an elaborate staging of imperial culture that corresponded with the celebration of Queen Victoria's jubilee'.[42] The Irish Exhibition, like the Indian Exhibition, provided viewers with an environment meant to replicate what might be seen by a tourist or traveller and, additionally, presented an 'idealized notion of a village community'.[43]

The contemporary press described the village and its inhabitants and while in London, although from different parts of Ireland, the artisans would have interacted with each other on a daily basis for the duration. Women worked together in the Embroiderers'

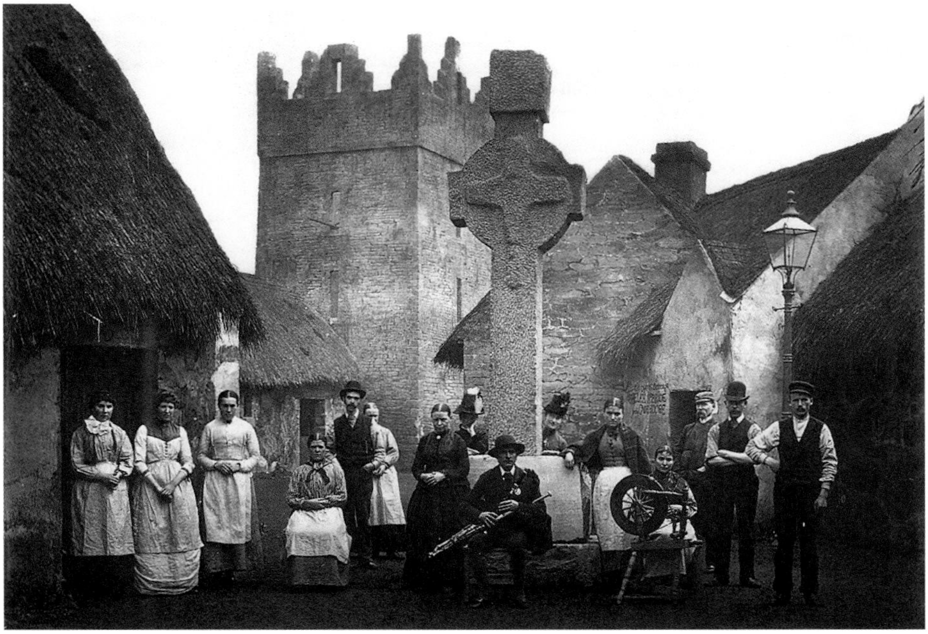

6.3 'View of the principal street in the Irish Village, Olympia, London', 1888.

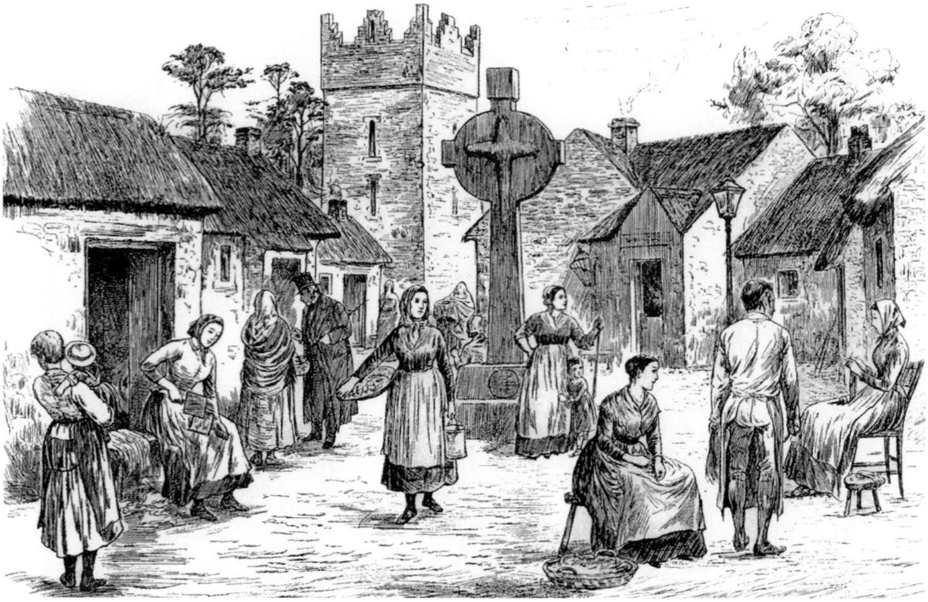

6.4 The Irish Village, Olympia, London, 1888. *Queen*, 7 July 1888, p. 8.

Cottage, the 'most extensive' exhibit at London's Olympia, among already-completed displays of '"Kells" portieres, curtains, and other decorative pieces, embroidered in Celtic designs, or in floral or geometrical patterns'; all the stitching was done on woollen or linen fabrics dyed with vegetable dyes, which were a 'speciality of the fund'[44] (see Figure 6.2). According to a lengthy article published in *Queen, The Lady's Newspaper*, the embroiderers were 'two of the Irish girls trained by' Una Taylor, an accomplished embroiderer who is best known for the stitching on the Irish National Banner, designed by Walter Crane and signed by Irish politician Charles Stewart Parnell, exhibited in the 1890 Arts and Crafts exhibition.[45] Shelah McBride from Gweedore, Donegal demonstrated plant dyeing from local plants of her district: 'from time to time she dyes in a little iron crock outside her cottage in the village street the wools required for use by the weaver in the adjoining cottage'.[46] Although she is not engaged in an activity in the photograph of the village 'inhabitants', a sketch reproduced in *Queen* shows two women 'dyeing wool' in front of the 'Kells Embroiderers' Cottage'; presumably one was Shelah McBride (Figure 6.4). McBride also demonstrated spinning of the 'wool that she has dyed' and when visitors went inside her cottage, they could view 'specimens of wool' of various colours hung from two frames.

How and where the villagers actually lived during the five months of the exhibition is unclear, but, because Alice Hart regularly trained Irish embroiderers in her storefront located on Wigmore Street in West London, she may have obtained regular accommodations for the participants. A sense of community probably prevailed even if that was a community of displacement; certainly, some of the participants obtained enough satisfaction to travel further afield with Alice Hart five years later when she took her Irish Village to Chicago.

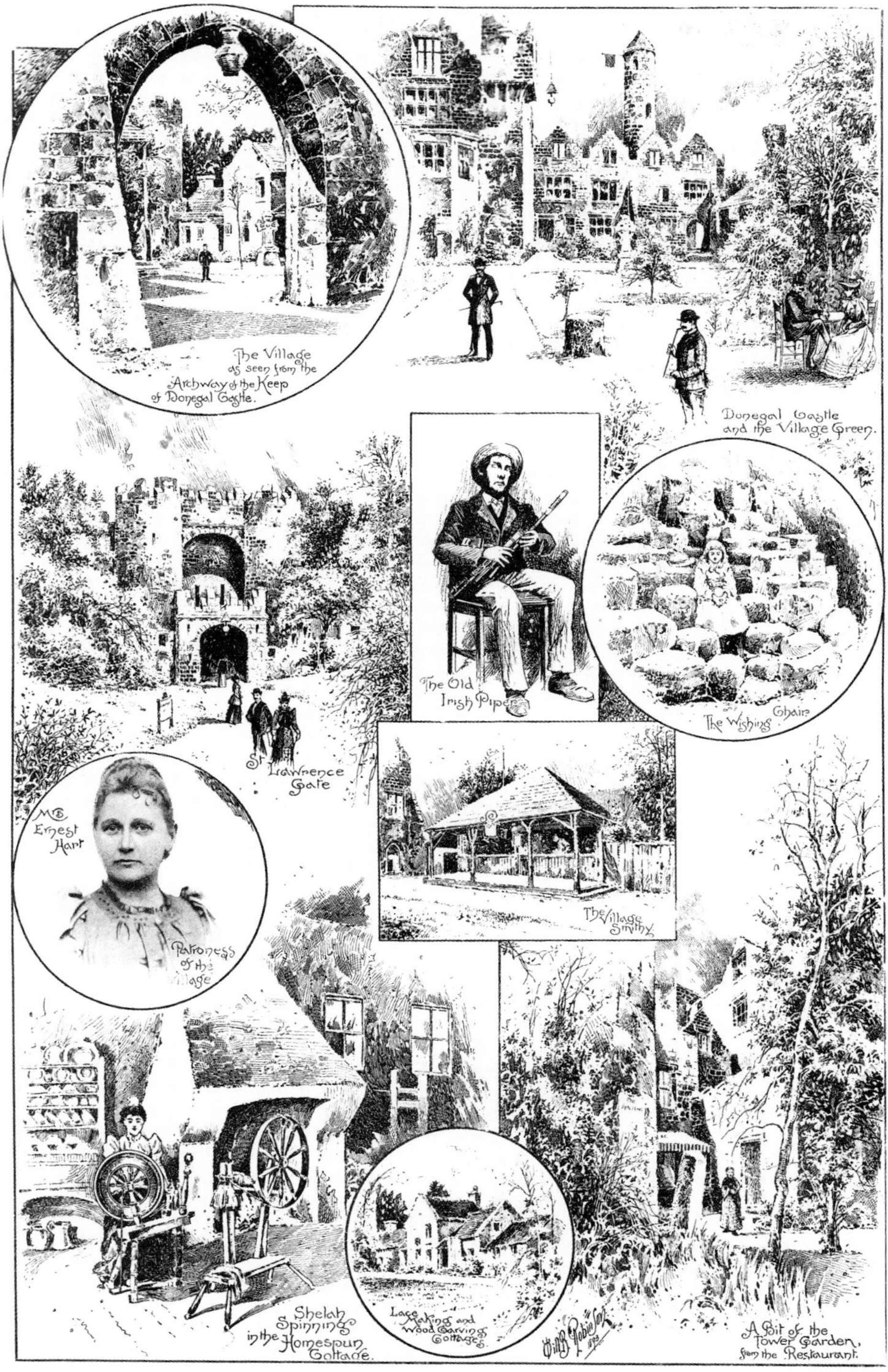

6.5 Hart's Irish Village, Chicago World's Fair, 1893. *Queen*, 14 October 1893, p. 636.

Hart arranged for the construction of a second Irish Village at the Chicago World's Fair of 1893 and, again, the artisans would have worked together for the six months of the fair.[47] The *Chicago Daily Tribune* announced the arrival of Charles McSweeney, 'the Piper of Donegal', along with other members of the 'fair community': Shelah McBride, the dyer and spinner (she had been in London in 1888), Hannah Gallagher, pillow lacemaker, Bridget Gildes and Rose Gildes, embroiderers, and Daniel O'Hara, the weaver, were all from Donegal (the names of their villages were not specified); Mary Mulligan, knitter, from Inniskillin; Mrs Glynn, lacemaker, from Limerick (also participated in the 1888 exhibition).[48] Albeit in a different context, Adam Arvidsson has argued that 'material goods can give a tangible materiality to … new imagined symbolic and social relationships'; that it may be possible 'to *fantasize realistically*, to produce new social identities and forms of community'.[49] One might imagine, then, that the production within the exhibition setting of 'Irish' products meant to advertise and promote Ireland and its industrious, thrifty and productive people may well have fostered a sense of community bolstered by the materiality of the objects (Figure 6.5).

The village plans were 'drawn by Prof. Hamlin, head of the Architectural Department of Columbia College, New York',[50] and it promised to be as exciting a spectacle for visitors as had the village built at Olympia five years earlier. Again, cottages were built that displayed the objects and the artisans, but there were additional features that signalled Hart's political affiliation with Ireland – something not present in the more restrained and ostensibly non-political spaces of the Olympia exhibition. Although she regularly and with enthusiasm expounded upon the quality of Ireland's handcrafted material culture and upon the dedication of Irish artisans, she never overtly mentioned Home Rule. She did, however, publically criticize the English for their policies in Ireland and she 'set to work to study the causes of Irish misery and disaffection'. In an 1885 speech she had acknowledged that she was 'always profoundly interested in the spectacle of a people stirred by their wretchedness into open rebellion, whether they be Bulgarians, Egyptians, or Irishmen',[51] which suggests that she supported independence for Ireland. At Chicago, her much-publicized unveiling of a statue of William Gladstone signalled her support for his liberal policy of Home Rule, which advocated limited or partial independence.

The United States was fruitful ground for the proclamation of support for Ireland whether for Home Rule or for more complete independence, and Hart's activities were championed by the American press and supported by the Catholic clergy. The nine-foot-high bronze replica statue of Gladstone was unveiled 'with much ceremony' near the end of June: the mayor of Chicago 'spoke of Mr Gladstone as the uncrowned King of the people of England and Ireland; Alice Hart "pulled the string that let the drapery fall"; a female singer in splendid voice sang "Come Back to Erin"'. A speech by Judge Moran followed in which he noted that Gladstone was then engaged in 'one of the most important political struggles of his life, in an effort to secure to the people of Ireland justice and home rule, and there is a peculiar fitness in the unveiling of his statue in America and in this village'. He referred to the respect awarded him by 'persons of Irish descent among our population'.[52] The speech was not one that would have been given in London in 1888 and indicates that while Hart claimed that her work in Ireland had 'no religious or political significance', her Irish Village suggested otherwise. In an interview she gave to the *Chicago Daily Tribune* before the opening of the Village she was quoted as saying: 'as early as 1879 and along to 1883 I felt that the people of Ireland were misunderstood and misrepresented by the English newspapers. I felt that England owed them a great deal'. She also insisted that her work was 'not a charity' – that if provided with opportunity, the craftspeople could become independent.[53]

Craft, Fair Trade and the Home Arts Movement

Could present-day analyses of global fair trade, particularly those affiliated with craft, given this discussion of one historical cottage craft organization, offer new insights into the home arts movement of the late nineteenth century? If one considers a global fair trade network as something other than strictly philanthropic, then an examination of the historical home arts movement could be considered as participating in a similar trajectory, one that is complex, multifaceted and diverse, a project that encouraged the development of technical skills that, in turn, could be put to use in a rapidly industrializing economy. A recent study suggests that there is 'substantial evidence' that there are 'wider benefits to society from increased skills' such as 'community cohesion and social capital'.[54] Perhaps the same is true historically; that is, enhanced technical skills and the production of material culture within a community may have led to more cohesion and a sense of 'social capital'. This slight shift in focus should lead to a re-examination and rewriting of the late nineteenth-century cottage craft movement as more than beneficent, not only revival, but also community based and interactive, a project that served artisans, consumers and patrons while at the same time signalling artisan independence, agency and nascent resistance.

Acknowledgments

The author gratefully acknowledges support from the Social Sciences and Humanities Research Council of Canada.

Sources Cited

Arvidsson, Adam, 'Brand Management and the Productivity of Consumption', in John Brewer and Frank Trentmann (eds), *Consuming Cultures, Global Perspectives: Historical Trajectories, Transnational Exchanges* (Oxford: Oxford University Press, 2006), pp. 71–94.

Beattie, Seán, 'Cottage Industries: Arts and Crafts in Donegal 1880–1920', *Donegal Annual*, 60 (2008), pp. 59–81.

Brison, Jeffrey, *Rockefeller, Carnegie, and Canada: American Philanthropy and the Arts and Letters in Canada* (Montreal and Kingston: Queen's University Press, 2006).

Crawford, Alan, 'United Kingdom: Origins and First Flowering', in Wendy Kaplan (ed.), *The Arts and Crafts Movement in Europe & America* (New York: Thames & Hudson, 2004), pp. 20–67.

Fridell, Gavin, 'The Fair Trade Network in Historical Perspective', *Canadian Journal of Development Studies*, XXV/3, (2004), pp. 411–28.

Gilbert, Helen and Chris Tiffin, *Burden or Benefit? Imperial Benevolence and Its Legacy* (Bloomington, IN: University of Indiana Press, 2008).

Greenhalgh, Paul, 'The Art and Industry of Mammon: International Exhibitions, 1851–1901', in John M. MacKenzie (ed.), *The Victorian Vision: Inventing New Britain* (London: V&A Publications, 2001), pp. 265–79.

de Groot, Joanna, 'Metropolitan desires and colonial connections: reflections on consumption and empire', in Catherine Hall and Sonya Rose (eds), *At Home With The Empire: Metropolitan Culture and the Imperial World* (Cambridge: Cambridge University Press, 2006), pp. 166–90.

Luddy, Maria, *Women and Philanthropy in Nineteenth-Century Ireland* (Cambridge: Cambridge University Press, 1995).

Mathur, Saloni, *India by Design: Colonial History and Cultural Display* (Berkeley, CA: University of California Press, 2007).

Milgram, Lynne B. and Kimberly M. Grimes (eds), *Artisans and Cooperatives: Developing Alternative Trade for the Global Economy* (Tucson, AZ: University of Arizona Press, 2000).

Moore, Geoff, 'The Fair Trade Movement: Parametres, Issues and Future Research', *Journal of Business Ethics*, 53/1–2 (2004), pp. 73–86.

O'Donoghue, D.J. *The Poets of Ireland* (Oxford: Oxford University Press 1912).

Russell Lynette (ed.), *Colonial Frontiers: Indigenous–European Encounters in Settler Societies* (Manchester: Manchester University Press, 2001).

Scrase, Timothy, 'Precarious production: globalisation and artisan labour in the Third World', *Third World Quarterly*, 24/3 (2003), pp. 449–61.

Peck, Amelia and Carol Irish, *Candace Wheeler: The Art and Enterprise of American Design, 1875–1900* (New York: The Metropolitan Museum of Art and Yale University Press, 2001).

Prochaska, F.K. *Women and Philanthropy in Nineteenth-Century England* (Oxford: Oxford University Press, 1980).

Notes

1 F.K. Prochaska, *Women and Philanthropy in Nineteenth-Century England* (Oxford, 1980), p. 5.

2 Maria Luddy, *Women and Philanthropy in Nineteenth-Century Ireland* (Cambridge, 1995), p. 3.

3 Gavin Fridell, 'The Fair Trade Network in Historical Perspective', *Canadian Journal of Development Studies*, XXV/3 (2004), p. 412.

4 Ibid., p. 413.

5 Geoff Moore, 'The Fair Trade Movement: Parametres, Issues and Future Research', *Journal of Business Ethics*, 53/1–2 (2004), p. 73.

6 FINE, as quoted in Moore, p. 73. 'FINE is an informal network that involves the Fairtrade Labelling Organizations International (**F**LO), the International Federation for Alternative Trade (**I**FAT), the Network of European Shops (**N**EWS) and the European Fair Trade Association (**E**FTA)'. Moore, 'The Fair Trade Movement', p. 85.

7 Sarah Lyon and Mark Moberg, *Fair Trade and Social Justice: Global Ethnographies* (New York, 2010), p. 7.

8 Lynne B. Milgram and Kimberly M. Grimes (eds), *Artisans and Cooperatives: Developing Alternative Trade for the Global Economy* (Tucson, 2000), p. 5.

9 Fridell, 'The Fair Trade Network in Historical Perspective', p. 412.

10 Timothy Scrase, 'Precarious production: globalisation and artisan labour in the Third World', *Third World Quarterly*, 24/3 (2003), p. 455.

11 Joanna de Groot, 'Metropolitan desires and colonial connections: reflections on consumption and empire', in Catherine Hall and Sonya Rose (eds), *At Home With The Empire: Metropolitan Culture and the Imperial World* (Cambridge, 2006), p. 168.

12 Seán Beattie, 'Cottage Industries: Arts and Crafts in Donegal, 1880–1920', *Donegal Annual*, 60 (2008), p. 78.

13 See, for example, Judith Hoag, *This Is Donegal Tweed* (Inver, Co. Donegal, 1987), and Beattie, 'Cottage Industries', pp. 59–81.

14 Alice Hart, 'The Cottage Industries of Ireland with an Account of the Work of the Donegal Industrial Fund', paper read before the Society of Arts, 11 May 1887, pp. 3–4. British Library Shelfmark: 8277 D 15. The Harts travelled to Ireland in the spring of 1883.

15 *The Times*, 8 December 1885, p. 4.

16 Ibid. All the homespuns, tweeds and woolens were 'manufactured by hand' and were either undyed or used vegetable dyes.

17 Amelia Peck and Carol Irish, *Candace Wheeler: The Art and Enterprise of American Design, 1875–1900* (New York, 2002), p. 48.

18 *New York Times*, 15 December 1886, p. 8.

19 In 1894, John Wanamaker provided space, rent free, in his Philadelphia department store in order to enable Hart to continue to exhibit in the United States when the Chicago World's Fair ended. *New York Times*, 21 February 1894, p. 2.

20 *The Times*, 27 November 1886, p. 6.

21 See http://www.officialroyalwedding2011.org/tag/homepage/page/1 [accessed 26 November 2011].

22 *Irish Times*, 26 May 1887, p. 6.

23 Moore, 'The Fair Trade Movement', p. 73.

24 Hart, 'The Cottage Industries of Ireland', p. 25.

25 *Chicago Daily Tribune*, 10 November 1892, p. 9.

26 *Illustrated London News*, 4 June 1887, p. 625.

27 *Queen*, 20 June 1885, p. 655. The exhibition celebrated inventions made since 1862.

28 Jeffrey Brison, *Rockefeller, Carnegie, and Canada: American Philanthropy and the Arts and Letters in Canada* (Montreal, 2006).

29 Helen Gilbert and Chris Tiffin, *Burden or Benefit? Imperial Benevolence and Its Legacy* (Bloomington, IN, 2008), p. 4.

30 Alan Crawford, 'United Kingdom: Origins and First Flowering', in Wendy Kaplan (ed.), *The Arts and Crafts Movement in Europe & America* (New York, 2004), p. 34.

31 Avril Denton used this phrase in her essay 'W.A.S. Benson: A Biography', in Ian Hamerton, *W.A.S. Benson: Arts and Crafts Luminary and Pioneer of Modern Design* (Woodbridge, Suffolk, 1988), p. 57.

32 Lynette Russell (ed.), 'Introduction', *Colonial Frontiers: Indigenous-European Encounters in Settler Societies* (Manchester, 2001), p. 1.

33 *The Lady*, 2 June 1887, p. 414 and *Lady's World*, 1887, p. 140.

34 Alice Hart, 'The Cottage Industries of Ireland', p. 18. The Countess Spencer, who later that year would host an exhibition in her home, was present at this talk.

35 *Queen*, 28 July 1888, p. 107.

36 *The Lady*, 2 June 1887, p. 414.

37 'Women's Industries at the Irish Exhibition', *Queen*, 11 August 1888, p. 168.

38 Timothy Scrase, 'Precarious production: globalisation and artisan labour in the Third World', *Third World Quarterly*, 24/3, (2003), p. 455.

39 Hart, The Cottage Industries of Ireland'.

40 The design and plan for the Irish Exhibition had been completed by May 1888. The Village, comprising 'a dozen cottages, built and thatched by native workmen in the Donegal fashion', was organized by the Earl of Leitrim and Alice Hart. *The Times*, 14 May 1888, p. 11. For a detailed discussion of the exhibition, see Janice Helland, 'Exhibiting Ireland: The Donegal Industrial Fund in London and Chicago', *Revue d'art canadienne/Canadian Art Review* (*RACAR*), XXIX/1–2 (2004), pp. 28–47. Caroline Rose Malloy also provided an excellent account of the 1888 Exhibition in 'How Ulster Became Irish: The 1924 British Empire Exhibition', paper presented at the annual conference of the Canadian Association for Irish Studies, Montreal, July 2011.

41 Paul Greenhalgh, 'The Art and Industry of Mammon: International Exhibitions, 1851–1901', in John M. MacKenzie (ed.), *The Victorian Vision: Inventing New Britain* (London, 2001), p. 266.

42 Saloni Mathur, *India by Design: Colonial History and Cultural Display* (Berkeley, 2007), p. 56.

43 Ibid., p. 57.

44 'The Exhibits of the Donegal Industrial Fund', *Queen*, 28 July 1888, p. 107.

45 Una Ashworth Taylor (b. 1857) was a close friend of arts and crafts designer May Morris. She was 'fervently Irish' and, in addition to her accomplishments as an embroiderer, also wrote poetry and prose 'almost exclusively on Irish and Catholic themes' for *United Ireland*, *Nation* and other journals. D.J. O'Donoghue, *The Poets of Ireland* (Oxford, 1912), p. 451.

46 *Queen*, 28 July 1888, p. 107.

47 The 1893 Chicago World's Fair was open from the beginning of May until the end of October.

48 The Chicago paper also gave the names of the blacksmith from Dublin, the linen weaver and the wood carver from Fanad. *Chicago Daily Tribune*, 4 May 1893, p. 2. The place-names were often misspelt and appear to have been written phonetically.

49 Adam Arvidsson, 'Brand Management and the Productivity of Consumption', in John Brewer and Frank Trentmann (eds), *Consuming Cultures, Global Perspectives: Historical Trajectories, Transnational Exchanges* (Oxford, 2006), p. 76.

50 *Chicago Daily Tribune*, 12 January 1893, p. 8. A.D.F. Hamlin (1855–1926) was professor of architecture at Columbia University and the author of *A Text-Book of the History of Architecture* (London, 1896). Of the buildings erected for the fair he wrote only: 'This was further shown [the excellence of American architects] in the buildings of the Columbian Exposition at Chicago in 1893. These, in spite of many faults of detail, constituted an aggregate of architectural splendor such as had never before been seen or been possible on this side the Atlantic. They further brought architecture into closer union with the allied arts and formed an object lesson in the value of appropriate landscape gardening as a setting to monumental structures' (p. 400).

51 'Cottage Industries and What They Can Do For Ireland. Being a verbatim report of an address given by Mrs Ernest Hart, at the Club House, Bedford Park, May 30th, 1885'. British Library shelfmark 8275 C 24. The Club House 'together with the School of Art opened in 1881' and was regarded as 'the centre of activity for the leading artistic and literary figures of the day', in *A Guide to Bedford Park: The first Garden Suburb*, prepared by Andrew Nunn and Associates. Available at: http://www.andrewnunnassociates.co.uk/public/files/17981E29-C2A1-411B-BA1B-D6FFF09DECC5/client_files/09691700-EA9F-4F6B-8345-48BD3585642E/pdf/guide.pdf [accessed 23 June 2012].

52 *Chicago Daily Tribune*, 26 June 1893, p. 2.

53 *Chicago Daily Tribune*, 20 November 1892, p. 26.

54 BIS: Department for Business Innovation and Skills, BIS Economics Paper No. 4, Supporting analysis for 'Skills for Growth: The national skills strategy', March 2010. Available at: http://www.bis.gov.uk/assets/biscore/economics-and-statistics/docs/10-604-bis-economics-paper-04 [accessed 17 June 2012].

7

Negotiating the Colonial Encounter: Making Objects for Export in the Andaman Islands, 1858–1920

Claire Wintle

Collections of non-European objects formed during the golden age of empire by Western missionaries, colonial officials or souvenir-hunting tourists have long been viewed as the material embodiment of wider imperial dispossession. In parallel to this, objects made for trade under these circumstances have been viewed as commodities that compromise the cultural integrity and economic independence of the producer. More recently, however, serious scholarly attention has been paid to the art commodities of marginalized and colonized peoples; objects made for export have been recognized for their importance in facilitating self-expression and community building.[1] Scholars of colonial collecting, for their part, have also worked to complicate the scholarship that assumes all collections formed under the auspices of colonialism are 'captives'.[2] This chapter sits within this revisionist movement, which seeks to deconstruct material production and collecting practices within the strictures of empire.

In a bid to stress the varied articulations of colonizing and counter-colonizing processes, this chapter investigates such processes as enacted on the Andaman Islands of India, an archipelago in the Bay of Bengal and the site of a major British penal colony from 1858. British contact with the 'Great Andamanese',[3] one of the four main delineations of indigenous peoples in the Andaman Islands, was a problematic encounter in imperial history. Individuals and communities were variously subject to a regime of pacification that included violent campaigns on distant villages and a programme that detained them in 'civilizing' residential institutions, or 'homes', at the heart of the British settlement. During this period the livelihoods of the Great Andamanese peoples were dramatically affected by an invasive policy of deforestation and the impact of imported diseases and narcotics. While the more persistently hostile and geographically distanced Ongee, Jarawa and Sentinelese communities largely escaped some of the most immediate results of British contact, the Great Andamanese, who occupied the lands in closest proximity to the British settlement, suffered disastrously. Once comprising ten distinct communities spread across South, Middle and North Andaman (three large islands that align to form the main skeleton of the archipelago), and estimated at 3,500 individuals in 1858, by 1901 these various peoples had been largely homogenized and amalgamated into the externally imposed category of 'Great Andamanese', and were recorded at 579.[4] Today, only 50 Great Andamanese people remain; they are kept in government care and have practically no memory of their languages or cultural histories;[5] Sita Venkateswar has described the British campaign as a form of ethnocide.[6]

However, while popular commentary and scholarship have consistently focused on the inevitability of Great Andamanese acquiescence under British force and rule,[7] and arguably failed to recognize the different stages of Great Andamanese resistance,[8] I argue that a study of the production and trade of indigenous objects in the Islands can help to nuance this understanding. Cultural practices bound up in ritual or lifestyle have only recently been considered as 'arenas of struggle',[9] but here material culture is seen as evidence of community strength and counter-hegemonic agency within wider colonial and postcolonial frameworks. To conceptualize the agency of some Andamanese producers and traders is not to deny the distorted power relations of the British–Andamanese encounter, or the long-term consequences that have threatened indigenous survival. This chapter will explore how objects made and traded through the framework of the Andaman homes were not substandard 'fakes' produced under subjugation, but that the very presence of these objects in the twenty-first century and the historical circumstances of their production constitute an everyday form of resistance.

Given the near extinction of the Great Andamanese today, the written and photographic colonial archive is the main available source for this investigation. There is no possibility of fully recovering the direct subaltern voice from the selective, ideological texts of elite domination, and, in the case of the Great Andamanese, very little information about named individuals' histories and opinions exists. Yet here, among a growing call for the recognition of indigenous agency in the Andaman and Nicobar Islands,[10] reading these sources 'against the grain' and strategically identifying a unified 'subaltern consciousness' is the only alternative to 'total unrepresentability'.[11] Examples of material culture produced by the Great Andamanese during the early twentieth century, since collected and deposited by various travellers in museums in Europe and North America, can also help to supplement this partial knowledge. As part of a growing recognition of the causal role of material culture in empire formation and development, scholars have come to advocate the particular use of object-centred research to reveal unique insights into imperial histories.[12] The facility to leave a material mark on the world is not subject to the same class, gender and ethnicity restrictions that dictate opportunities to contribute to written colonial archives or published documents, and accordingly the study of objects has the potential to produce insights into the lives of those peoples who are excluded from such modes of representation. Clearly, however, any story of 'Great Andamanese' activity, agency and community, particularly given the disparate origins of this group, can be only a partial one.

The Andaman 'Homes' and Objects for Trade

The British arrived in the Andaman Islands in 1858, driven by the need to incarcerate participants in the Indian Mutiny/Rebellion of 1857, and the longer-term desire to quash the ongoing threat posed by the region's hostile indigenous populations to shipwrecked crews and refuelling traders.[13] The Andamanese had long been seen by European observers as 'a race of men the least civilized, perhaps, in the world'.[14] But following colonization, reports and examples of Britain's new subjects, with their 'African' physiognomy and 'simple' material culture, became highly sought after as verification of socio-evolutionary theories. By the 1870s, the demands of scientists, colonial officials and other visitors seeking a souvenir of their travels created a lively trade in Andamanese material culture, and led to the production of objects for financial reward: these were made and sold from the government-run Andaman homes located around South Andaman Island.

Originally founded in July 1863 as a 'small house' in which four captured Andamanese were 'watched by a strong guard of Natives',[15] the government homes soon became a network of stations located at over 16 different sites around South Andaman. While the various homes differed somewhat in longevity, size and purpose, the main home at Haddo averaged about 95 residents in the 1880s,[16] its central purpose to manage the Andamanese and utilize their skills for the benefit of the colony. Broadly, the Andaman homes fit into the wider template of residential institutions commonly set up in imperial situations across the globe with the aim of training and controlling groups seen as in need of correction: British policy towards the Andamanese changed throughout the nineteenth and twentieth century,[17] but, generally, inmates were attracted to the homes through bribes and occasionally by force, and once in residence, subject to a series of controlling mechanisms designed to 'civilize' and 'pacify' them. In the 1870s and 1880s, as with similar institutions in North America and Australia, inmates were required to wear Western clothes, learn the English language, and adopt a grain diet in place of the usual pig and fish, although this 'corrective' framework became more informal later in the century.[18]

Throughout, 'constructive' employment was consistently considered to be one of the most useful tools in the civilizing mission linked with the homes.[19] The Andamanese were encouraged to collect and sell shells, eggs and honey, produce thatching leaves, and make examples of their own material manufactures for sale. In January 1875, a shop 'for the sale of their commodities' was established, and although this specific site was shut later that year in favour of an enduring 'system of house to house sales', an institution described as a 'little semi-shop, semi-museum' selling bows and arrows and body adornments was still trading in 1909.[20] In later years, visitors to the homes also found the Andamanese ready to perform their ceremonial dances upon request.[21] Most of the income from these sales and performances went directly to the 'credit of the Government', and more specifically, to the maintenance of the homes.[22] Objects and services were also exchanged for goods such as clay pipes, scrap iron, textiles, beads, mirrors, sugar, grain and cooking utensils. The value of some of these commodities, such as tobacco and alcohol, relied on dependencies purposely fostered by the British, who stood to gain from regular contact with Andamanese who came to the settlements seeking to fuel their addictions.

Given the context of their production and some of the imported commodities exchanged, these objects have generally been framed as the unnatural product of forced labour.[23] Certainly, the manufacture of this strand of Andamanese material culture was part of a wider employment encouraged and regulated by British officers, and, in the earlier years of the system, enforced by police guards or prisoner-overseers from the penal settlement.[24] By 1900, residencies at the homes were more fluid, and the Andamanese seemed to come and go subject to their own requirements, visiting at certain times and removing themselves from British control and influence at others. Indeed, as we shall see, British control was haphazard, and the Andamanese retained their own cultural practices and social arrangements in complex ways. Despite the inconsistencies in the ways in which the homes were managed, however, it is clear that at certain stages Andamanese people were forcibly detained, and, through a campaign of persuasion, pressure and forced addiction, compelled to engage in a general labour that included the production of their material culture.

As a stable context for intertribal communication, and the breeding ground for infectious disease, the homes have also been widely recognized as the principal site for the homogenization and transformation of the indigenous population. Extreme demographic change and particular cultural permutations have certainly been the long-term result of the atypical situation they facilitated – particularly among the various communities comprising the Great Andamanese, social delineations and modes of material production became 'thoroughly mixed up'.[25] The extent of this amalgamation of both social divisions and creative styles is suggested in the grave observation that by 1899 the only Great Andamanese encampments on South Andaman were in the homes.[26] By the turn of the century much Great Andamanese material culture was seemingly produced either within these institutions, or within a cultural framework subject to their influence.

However, while Vishvajit Pandya has suggested that the homes resulted in the loss of Andamanese dignity and identity,[27] and Satadru Sen has argued that they stripped the Andamanese 'of their right to represent themselves',[28] here it is suggested that the manufacture and sale of Andamanese material objects to a non-indigenous market can be recognized as a creative and positive response to new requirements, requests and impositions from external sources. David Tomas has identified the objects produced for sale in the homes as '"curios" of nominal touristic value', and as 'reproductions', which are different from 'the real thing',[29] but it could also be argued that these continue to be objects that built on a distinctive Andamanese worldview. In many respects, to draw on Ruth Phillips's work on souvenir wares, these can be termed 'autoethnographic expressions', which represent their makers, even while engaging with externally imposed terms.[30] The manufacture and sale of body adornments made by Great Andamanese women during the late nineteenth and early twentieth centuries can serve as a useful example to illustrate these arguments.

Body Adornment and Identity

The everyday dress of Great Andamanese women consisted of a mandatory four or five waist adornments, formed of layered pandanus-leaf strips looped six times and secured with twine. For married women, additional tassels of pandanus-leaf ribbons were attached and positioned towards the rear. Shell necklaces and bands of twine or pandanus leaf were more sporadically worn around the wrists, arms, legs and ankles according to personal choice (see Figure 7.1).[31] The lowest of these belts would have secured a number of loose, layered leaves positioned in order to cover the pubic area, and each would have been decorated, most commonly with fine cross-hatched markings incised with a sharpened shell. Such objects were also frequently decorated with white and red clay, and strings of dentilium shell. Clay and shell decoration were used to express style and creativity, but they more often implied the ceremonial nature of the object, and were incorporated for their perceived capacity to help the wearer manage their weight, temperature and smell, as each of these variable states had a specific purpose in attracting or dispelling the spirits that surrounded

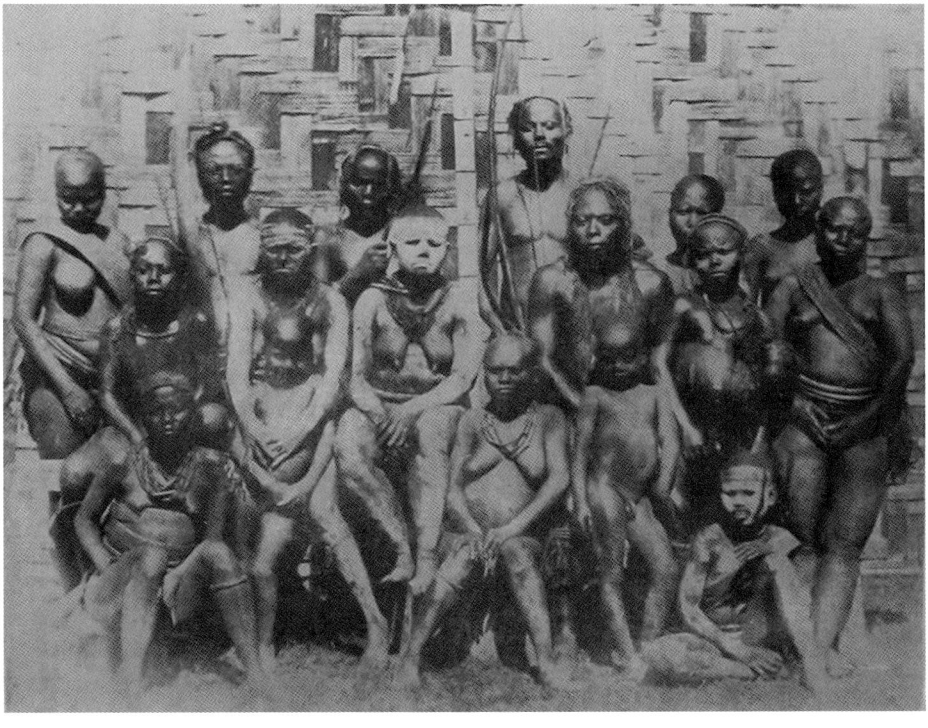

7.1 Edward Horace Man. Photograph of group of Andamanese men and women, probably Aka-Bea from Rutland Island, at Port Blair, 1875 or 1876. From E.H. Man, *A Dictionary of the South Andaman (Aka-Bea) Language* (Bombay: British India Press, 1923), reprinted from the *Indian Antiquary*, Plate XIII.

The male in the centre of the back row is recognizable from other photographs as Maia Biala, chief of Rutland Island.

the Andamanese.[32] Traditionally, each woman made her own adornments, and those required by her husband.[33]

Comparative analysis of objects and photographs collected in the 1870s and in the first two decades of the twentieth century demonstrates some important elements of continuity and creativity among objects and makers in the fact of encroachment from outsiders. Figure 7.1 is an early example of the 'scientific' imagery produced for the benefit of the anthropological community in Europe, as the orderly arrangement of the sitters and central measuring stick indicate. The photographer, Edward Horace Man, was a colonial official who had brokered a tense but fairly sustained relationship with some of the communities of South Andaman both in and away from the homes, often using his contacts as subjects for images that aimed to demonstrate physical type and cultural practices. However, in comparison with other examples of his photography, which carefully reconstructed scenes of everyday activities and relied on regular, long-term residents of the homes to comply with these demands,[34] this seems to be one of his more informal images: these are not the 'inmates of the homes' referred to elsewhere, but appear instead to be passing visitors from Rutland Island (to the south of South Andaman), and as such comprise a group (and their adornments) relatively unaffected by the disciplining mechanisms and performative nature of the homes.

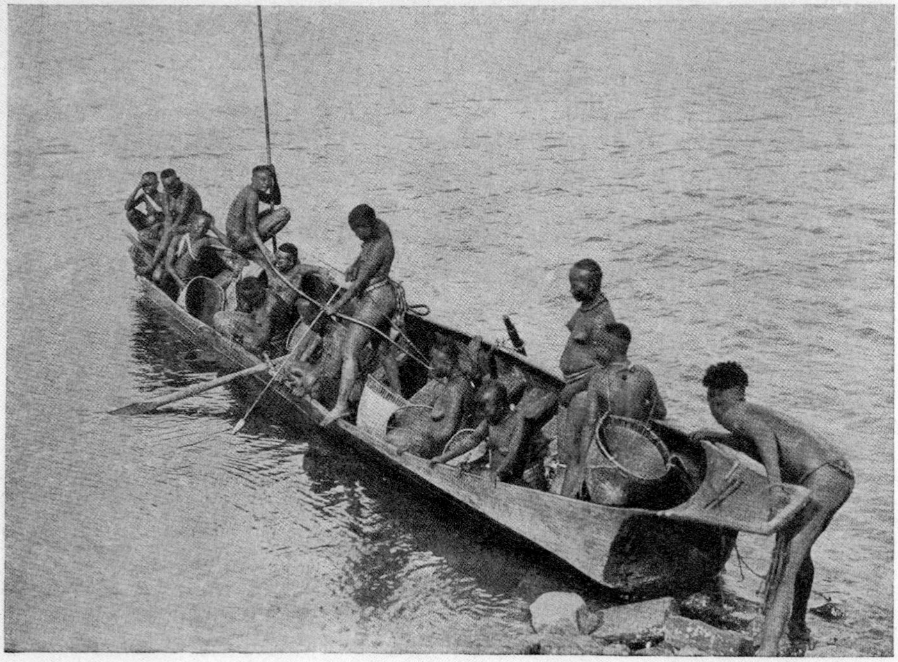

A FISHING-PARTY

7.2 Photograph of Andamanese men and women 'fishing' near the Andaman home at Haddo, c. 1911.

Figures 7.2 and 7.3 are much later images: the former shows a carefully staged group of men and women based at the home, 'fishing' for the benefit of Frederick Taylor, an amateur explorer who briefly visited South Andaman in about 1911. Figure 7.3 depicts three Andamanese women and their offspring, probably in late 1914, also posing for the camera, but this time the composition appears more hastily constructed. Included in a personal photograph album that belonged to Mrs W.H. Burt, the wife of a junior army officer stationed on the Islands in 1914 and 1915, the image was probably taken by Burt or a companion, and could have been photographed in the main home at Haddo, which she is known to have visited.[35] Given the highly constructed nature of Figure 7.2 and the more impromptu appearance of Figure 7.3, it may be the case that the Andamanese in the former image are responding to a perceived or actual demand from Taylor to dress in 'traditional' or 'authentic' clothes for the benefit of a Western audience.[36] This suggestion is supported by the presence of the length of cotton worn by the woman in the centre of Figure 7.3;[37] here, away from the constructed performance of 'being Andamanese', the image seems to capture some of the outside influence and changes incorporated into the dress of Andamanese women by 1914. Both images are useful for our purpose, however, because they depict adornments made for use by the Andamanese in their everyday life (Figure 7.3) and adornments ostensibly made for ceremony, performance and sale (Figure 7.2).

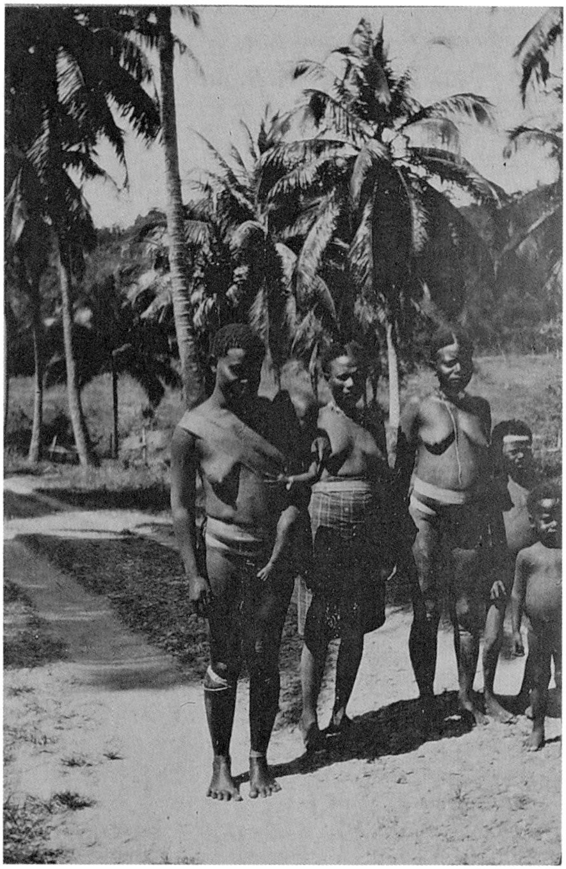

7.3 Photograph of Great Andamanese women, with children, probably in the Andaman home at Haddo, c. 1914.

Notwithstanding their traces of a Western presence, these photographs, along with contemporary written descriptions and objects now in museum collections, do not depict dramatic shifts in the physical construction of the body adornment of Great Andamanese women between 1875 and 1914. The clay body paint seen in Figure 7.1 but not in the later images (probably worn in this case as a sign of mourning) was observed elsewhere by visitors to the Islands in the early twentieth century. The material

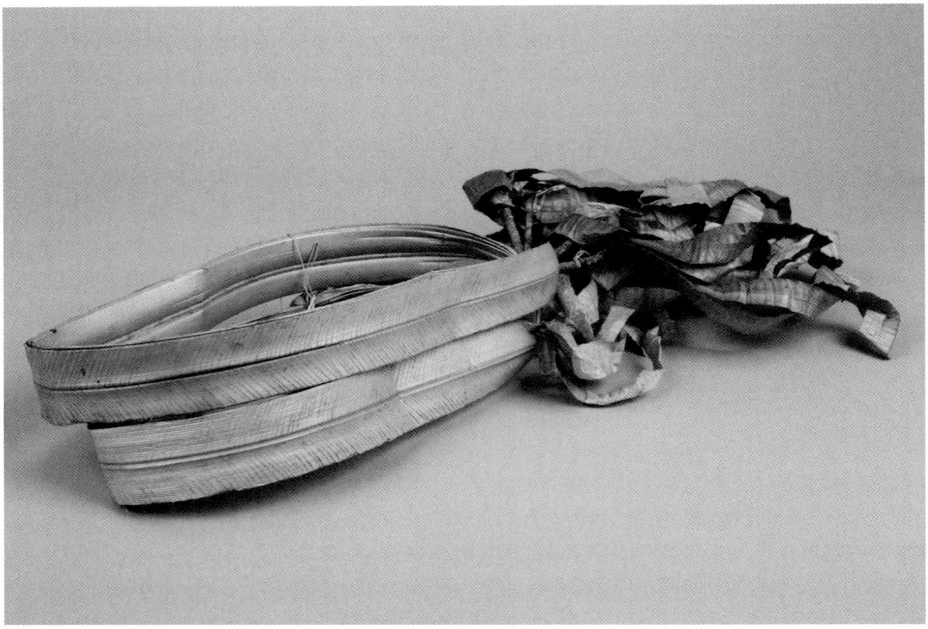

7.4 Waistband of three loops of pandanus leaf, with incisions, Great Andaman, c. 1900.

ornaments worn in Figures 7.1 and 7.2 are physically identical,[38] and Figure 7.3 also shows the standard everyday body adornments described above, although notably these seem to be made of finer and slightly wider bands of pandanus leaf, and create a smoother line with the contours of the body than some other examples depicted elsewhere. A comparable waistband in the collection of Royal Pavilion and Museum, Brighton & Hove, probably collected around 1900 (Figure 7.4), is of a similarly fine construction and made of only three loops of pandanus leaf, but still decorated with fine incised lines in keeping with earlier examples. Similarly, with ceremonial adornments collected by Taylor, Burt and other visitors to the Islands after the turn of the century, strings of dentalium shell, clay decorations and incisions are also still present, but occasionally supplemented by blue and turquoise glass beads procured from exchanges with Europeans, and lengths of red cloth.[39]

The implications of these physical changes will be examined shortly, but the metaphysical implications of the changes in object production and use value facilitated by the homes must also be introduced here: to produce an object for personal use, or for a close relative, has very different implications from making an identical object for prospective trade.[40] Notably, the Andamanese did produce material culture for reasons other than personal use: gift-exchange was an integral part of the interaction among Andamanese individuals and communities,[41] and while the relationship of exchange as forged between indigenous groups and outsiders was not constructed in the same framework as the demonstrative gift-exchange engaged in between the Andamanese,[42] trade had always been implicated in Andamanese production. Importantly, it is unlikely that Great Andamanese material, whether for everyday use or ceremonial purposes, was ever mass-produced for a large market. A statement of the earnings of the Andamanese attached to the homes between 1876 and 1882 demonstrates that the sale of 'bows, arrows, &c'. made an average of 125 rupees out of a total average income from other activities of 1724 rupees per annum.[43] While there are no complete statistics for later years, the monthly reports for the Andaman homes after 1882 do not paint a picture of constant industry and forced labour, but rather present the homes as spaces of sporadic production. Particular objects were available on commission for both individual travellers and the government,[44] but tourists did not visit the Islands en masse, and while most major ethnographic collections in Europe contain Andamanese material forwarded by enthusiastic colonial officials, these tend to contain one or two examples of each 'specimen' of Andamanese material culture. Objects were speculatively produced and stored in small numbers to meet this demand,[45] but the production of Andamanese material culture for an outside audience cannot be said to have moved to a stage where desire for profit or economic necessity superseded long-term aesthetic and cultural standards of manufacture.[46]

Nevertheless, while objects did retain an inherent material structure and continued to perform traditional roles in Andamanese life, physical and conceptual changes had certainly been enacted as a result of European impositions. Such continuities and modifications are significant for our understanding of how the Great Andamanese managed their sense of community and identity within a British regime that sought to annihilate elements of indigenous culture. At a basic level, if creativity is identified as one of the most potent attributes of humanity, these crafted objects can be seen as markers of Andamanese life and as affirmation of their producers' talents, traditions and histories.[47] Even in the homes, creating objects would have required the maker to participate in active processes of selection, experimentation, disposal, discovery and problem-solving that required individual input and thought. As sociologists have begun to make clear, 'making is thinking',[48] and while employment demands made of the homes' inmates and the conditions of British collecting conventions may have formed the catalysts for the production of some objects, those external forces should not be credited with all the commitment, imagination and creativity inherent in their development.

Moreover, despite external influences, these objects were firmly rooted in Andamanese design histories and worldviews. Whether made for personal use or for trade, the production of Andamanese material culture was created through sequences of operation that were socially learnt and culturally embedded. The manufacture of body adornments, for example, would continue to be rooted in a worldview that saw the painting and cutting of pandanus-leaf belts as gendered acts that ensured the wearer's safety and assistance from the spirits and worked to cement community membership.[49] Even in later examples, where some belts show a less robust construction, these culturally specific acts of binding and incision remain integral to the object (see Figure 7.4). Great Andamanese creation myths about the first man and woman incorporate material culture as signifiers of gender;[50] these perceptions, and the significance of binding and tying, valued as acts that impact upon spirit–human relations through transforming the state of the human body, would also have continued to influence the making process.[51] Richard Sennett and Christopher Tilley have argued that through systems of craft and manufacture, people both construct and comprehend their social selves:[52] an alternative final purpose would not necessarily have erased these culturally situated, long-term practices.

Making and Community in the Andaman Islands

The homes were undoubtedly British institutions, infiltrated by the political, social and economic demands of outsiders, but the very presence of the indigenous residents and the cultural processes enacted there also ensured that the homes had identities as Andamanese spaces. As suggested, by 1900 the homes were more transient arenas through which the Andamanese seemed to move at will. But even in the earlier stages of the British regime, when inmates were subject to greater restrictions, the Andamanese chose to modify the apparatus of British control in accordance with their own routines, and, as colonial officials acknowledged, 'always built their own huts ... inside'.[53] In 1869, a visitor to one home described his encounter with a woman who 'possessed great influence over the tribe, and acted as arbitrator in all disputes',[54] hinting at the ways in which established indigenous systems of community mediation continued to retain effect in conjunction with British structures of authority.[55] He also described how the homes' residents engaged in long-standing methods of personal hygiene, cooking and mourning, and how, as already suggested, they continued to produce and use the items of material culture traditionally associated with these processes.[56]

Recent scholarship has come to see the process of making as related to community-building and cultural cohesiveness; it often involves a social dimension, requiring the sharing of skills across generations, and utilizing other societal structures.[57] In the Andaman Islands, parents educated and socialized their children and foster children through teaching them how to make and use certain items of material culture according to appropriate gender roles: boys began to learn the occupations of men through the use of toy bows and arrows made by their elders, while girls were expected to assist their mothers in gathering shells and materials for their adornments.[58] Crafting in the Andaman homes continued to build on social relations among the Andamanese; most raw materials would still have been collected together and traded between communities, while sitting and making adornments, for example, would still have provided a time for women to come together and discuss their lives. The claim that such objects represent the British desire to instil a work ethic in the Andamanese is also a fallacy; laziness had always been regarded as highly anti-social among the Andamanese, and much of the working day for women, even before British influence, was taken up by making adornments and other implements used in cooking and collecting food.[59]

Moreover, where new stylistic and functional strains of Andamanese material culture did emerge from the new inter-community networks facilitated by British intervention, or in reaction to the demands of outsiders, they should not be dismissed as 'simulated elements'[60] of Andamanese culture. To reject those aspects of Andamanese production that diverge from the externally identified 'traditional' standard as 'inauthentic' is to continue to place the Andamanese in the state of arrested evolutionary development devised by nineteenth-century anthropology. Despite the continued use of descriptive monikers that dub the Andamanese 'Stone Age Islanders',[61] their cultural development has never been stagnant: the use of salvaged iron as a replacement for bone in the manufacture of tools and equipment during the nineteenth century is but one material instance of the adaptive nature of Andamanese society. In 1906–8, anthropologist A.R. Radcliffe-Brown observed the diversity in the design of body painting, for example, and the competition among women to produce novelty and innovation in their work.[62] As Pandya notes in his fascinating study of the complex ways in which the Andamanese continue to incorporate non-indigenous textiles into otherwise organically formed body adornments, shifts in object production have always been an important part of negotiating changes in Andamanese culture.[63] The same objects that were sold to outsiders appear to have been worn by their makers, suggesting that the habitual creative reordering of Great Andamanese objects, such as in, for example, the incorporation of blue beads in body adornments, can also be read as an indigenous-led desire to explore new fashions, materials and techniques. The production of 'tourist art' within the homes can thus be seen as an instance of Andamanese vitality and fluidity, here in the face of British subjugation.

It was in the interests of government officials who produced the colonial archives to silence Andamanese agency and dissent: their job was to pacify their charges and report on their progress in this matter to superiors on the mainland. Today, an alternative agenda seeking to redress past historical wrongs has accidently silenced the subjugated in order to deride the colonizer. However, recent scholarship has pointed to the fragile grip the British actually had on the Andamanese in the late nineteenth century,[64] and this chapter has highlighted the continued if cautious dynamism of Great Andamanese culture well into the twentieth century, even after the homes laid claim in some way to much daily life on South Andaman. The under-researched material remnants of this later period illustrate the continuation and creative adaptation of a culture that was firmly rooted in Andamanese identity and community. Despite outside influences, these were Andamanese objects, built on an indigenous design history, social order and worldviews; in making these objects, the Great Andamanese enacted processes of socialization, inter-community trade and interaction.

'Agency' is a complex phenomenon, and it is difficult to argue that in making objects the Great Andamanese consciously shifted the course of action instigated by the British and militated against the long-term impact of outsider encroachment on indigenous existence.[65] There is also a danger of relying upon a romantic, Ruskinian vision of the handcrafted object as the keystone of community, and, of course, a problem with the assumption that shared experiences of production necessarily result in shared subject positions.[66] But the objects produced are evidence of a form of cultural cohesion under subjugation and physical confirmation that the Great Andamanese had not submitted to a British 'civilizing' programme by 1920.[67] They make a pragmatic difference to the historical record and demonstrate Great Andamanese capacity to act against cultural suppression.

Sources Cited

Abbi, Anvita, *Endangered Languages of the Andaman Islands* (Munich: Lincom Europa GMBH, 2006).

Adamson, Glenn, 'Craft in Action: Life, Art, Design: Section Introduction', in Glenn Adamson (ed.), *The Craft Reader* (Oxford: Berg, 2010), pp. 457–63.

Anderson, Clare, *The Indian Uprising of 1857–8: Prisons, Prisoners and Rebellion* (London: Anthem Press, 2007).

Anderson, Clare, 'Colonization, Kidnap and Confinement in the Andamans Penal Colony, 1771–1864', *Journal of Historical Geography*, 37 (2011), pp. 68–81.

Ball, Valentine, 'Visit to the Andamanese "Home," Port Blair, Andaman Islands', *Indian Antiquary*, 3 (1874), pp. 171–3.

Barker, Chris, *The Sage Dictionary of Cultural Studies* (London: Sage, 2004).

Burt, W.H., untitled typescript detailing life on the Andaman Islands, undated [1914], Department of Africa, Oceania and the Americas' Archive (Eth Doc 1714), British Museum.

Clifton, Mrs [Violet] Talbot, *Pilgrim to the Isles of Penance: Orchid Gathering in the East* (London: John Long Limited, 1911).

Dutta, Pratap C., *The Great Andamanese: Past and Present* (Calcutta: Anthropological Survey of India, 1978).

Edwards, Elizabeth, 'Science Visualized: E.H. Man in the Andaman Islands', in Elizabeth Edwards (ed.), *Anthropology and Photography 1860–1920* (London: Yale University Press and Royal Anthropological Society, 1992), pp. 108–21.

Gauntlett, David, *Creative Explorations: New Approaches to Identities and Audiences* (Abingdon: Routledge, 2007).

Gauntlett, David, *Making Is Connecting: The Social Meaning of Creativity, from DIY and Knitting to Youtube and Web 2.0* (Cambridge: Polity, 2011).

Gosden, Chris, and Chantal Knowles, *Collecting Colonialism: Material Culture and Colonial Change* (Oxford: Berg, 2001).

Graburn, Nelson H.H., 'Introduction: The Arts of the Fourth World', in Nelson H.H. Graburn (ed.), *Ethnic and Tourist Arts: Cultural Expressions from the Fourth World* (Berkeley: University of California Press, 1976), pp. 1–32.

Guha, Ranajit, *Elementary Aspects of Peasant Insurgency in Colonial India* (New Delhi: Oxford University Press, 1983).

Harrison, Andrew, *Making and Thinking: A Study of Intelligent Activities* (Brighton: Harvester Press, 1978).

Haynes, Douglas and Gyan Prakash, 'Introduction: The Entanglement of Power and Resistance', in Douglas Haynes and Gyan Prakash (eds), *Contesting Power: Resistance and Everyday Social Relations in South Asia* (New Delhi: Oxford University Press, 1991), pp. 1–22.

Man, Edward Horace, 'On the Aboriginal Inhabitants of the Andaman Islands (Part III)', *Journal of the Anthropological Institute*, 12 (1883), pp. 327–434.

Man, Edward Horace, *A Dictionary of the South Andaman (Aka-Bea) Language* (Bombay: British India Press, 1923), reprinted from the *Indian Antiquary*.

Man, Edward Horace, *On the Aboriginal Inhabitants of the Andaman Islands with a Report into the Languages of the South Andaman Island by A.J. Ellis* (London: Royal Anthropological Institute, 1932).

McDougall, Dan, 'How a "Stone Age" Tribe Is Keeping the Modern World at Bay', *The Week*, 18 February 2006, pp. 44–5.

Mukerjee, Madhusree, *The Land of Naked People: Encounters with Stone Age Islanders* (London: Houghton Mifflin, 2003).

O'Hanlon, Michael and Robert L. Welsch (eds), *Hunting the Gatherers: Ethnographic Collectors, Agents and Agency in Melanesia, 1870s–1930s* (New York: Berghahn, 2000).

Pandya, Vishvajit, *Above the Forest: A Study of Andamanese Ethnoanemology, Cosmology, and the Power of Ritual* (New Delhi: Oxford University Press, 1993).

Pandya, Vishvajit, 'The Andaman Islanders of the Bay of Bengal', in Richard B. Lee and Richard Heywood Daly (eds), *The Cambridge Encyclopedia of Hunters and Gatherers* (Cambridge: Cambridge University Press, 1999), pp. 243–7.

Pandya, Vishvajit, 'Deforesting Amongst Andamanese Children: Political Economy and History of Schooling', in Barry S. Hewlett and Michael E. Lamb (eds), *Hunter Gatherer Childhoods: Evolutionary, Developmental, and Cultural Perspectives* (New Brunswick, NJ: Aldine Transaction, 2005), pp. 385–406.

Pandya, Vishvajit, *In the Forest: Visual and Material Worlds of Andamanese History (1858–2006)* (Lanham, MD: University Press of America, 2009).

Phillips, Ruth B. and Christopher Steiner (eds), *Unpacking Culture: Art and Commodity in Colonial and Postcolonial Worlds* (Berkeley, CA: University of California Press, 1999).

Phillips, Ruth B., *Trading Identities: The Souvenir in Native North American Art from the Northeast, 1700–1900* (Seattle, WA and Montreal: University of Washington Press and McGill-Queen's University Press, 1998).

Portman, Maurice Vidal, *A History of Our Relations with the Andamanese: Compiled from Histories and Travels, and from the Records of the Government of India,* 2 vols (Calcutta: Office of the Superintendent of Government Printing, 1899).

Pratt, Mary Louise, *Imperial Eyes: Travel Writing and Transculturation* (London: Routledge, 1992).

Proceedings of the Revenue and Agricultural Department, Asia, Pacific and Africa Collections, British Library.

Proceedings of the Superintendent of Port Blair and the Nicobars, Asia, Pacific and Africa Collections, British Library.

Sen, Satadru, 'Policing the Savage: Segregation, Labor and State Medicine in the Andamans', *Journal of Asian Studies*, 58/3 (1999), pp. 753–73.

Sen, Satadru, *Disciplining Punishment: Colonialism and Convict Society in the Andaman Islands* (Oxford: Oxford University Press, 2000).

Sen, Satadru, *Savagery and Colonialism in the Indian Ocean: Power, Pleasure and the Andaman Islanders* (London: Routledge, 2010).

Sennett, Richard, *The Craftsman* (London: Allen Lane, 2008).

Spivak, Gayatri Chakravorty, *In Other Worlds: Essays in Cultural Politics* (New York: Routledge Classics, [1997] 2006).

Taylor, Frederick, 'Native Life in the Andaman Islands: A British Penal Settlement for India, with Pictures from Photographs Taken for the Writer', *Century Illustrated Monthly Magazine*, 82/6 (1911), pp. 891–904.

Temple, Richard Carnac, 'Andaman and Nicobar Islands: Report on the Census', in *Census of India, 1901* (Calcutta: Office of the Superintendent of Government Printing, 1903), pp. 1–33.

Temple, Richard Carnac, *Remarks on the Andaman Islanders and Their Country* (Bombay: British India Press, 1930).

Thomson, Allen, 'Description of Andamanese Bone Necklaces', *Journal of the Anthropological Institute of Great Britain and Ireland*, 11 (1881), pp. 295–309.

Tilley, Christopher, 'Introduction', in Christopher Tilley, Webb Keane, Susan Kuechler, Mike Rowlands and Patricia Spyer (eds), *Handbook of Material Culture* (London: Sage, 2006), pp. 1–6.

Tomas, David, 'An Ethnography of the Eye: Authority, Observation, and Photography in the Context of British Anthropology 1839–1990' (unpublished doctoral thesis, Department of Anthropology, McGill University, 1987).

Van Dommelen, Peter, 'Colonial Matters: Material Culture and Postcolonial Theory in Colonial Situations', in Christopher Tilley et al. (ed.), *Handbook of Material Culture* (London: Sage, 2006), pp. 104–24.

Venkateswar, Sita, *Development and Ethnocide: Colonial Practices in the Andamans* (Copenhagen: International Work Group for Indigenous Affairs, 2004).

Wintle, Claire, 'Objects, Images, Imaginings: New Perspectives on the Material and Visual Culture of the Andaman and Nicobar Islands', *Journal of Museum Ethnography*, 20 (2008), pp. 145–52.

Wintle, Claire, *Colonial Collecting and Display: Encounters with Material Culture from the Andaman and Nicobar Islands* (Oxford: Berghahn, 2013).

Notes

Some of the material and ideas discussed here are based on discussions first put forward in my *Colonial Collecting and Display: Encounters through Material Culture in the Andaman and Nicobar Islands* (Oxford, 2013).

1 For example, Ruth Phillips and Christopher Steiner (eds), *Unpacking Culture: Art and Commodity in Colonial and Postcolonial Worlds* (Berkeley, 1999). See also Anne de Stecher, Chapter 2, this volume.

2 For example, Michael O'Hanlon and Robert Welsch (eds), *Hunting the Gatherers: Ethnographic Collectors, Agents and Agency in Melanesia, 1870s–1930s* (New York, 2000).

3 The category 'Great Andamanese' is an externally imposed construct applied by the British to ten discrete indigenous groups who occupied South, Middle and North Andaman in 1858 (Aka-Cari, Aka-Kora, Aka-Bo, Aka-Jeru, Aka-Kede, Aka-Bol, Oko-Juwoi, A-Pucikwar, Akar-Bale, Aka-Bea). Given the complexity of the social structure of these communities and particularly the shifts that occurred as a result of the British presence, I use the term 'Great Andamanese' here for ease of reference, but this is acknowledged as problematic in many respects.

4 Richard Carnac Temple, 'Andaman and Nicobar Islands: Report on the Census', in *Census of India, 1901* (Calcutta, 1903), pp. 10 and 30.

5 As of 2005, cited in Anvita Abbi, *Endangered Languages of the Andaman Islands* (Munich, 2006), pp. i and 20. Vishvajit Pandya, 'The Andaman Islanders of the Bay of Bengal', in Richard B. Lee and Richard Heywood Daly (eds), *The Cambridge Encyclopedia of Hunters and Gatherers* (Cambridge, 1999), p. 244.

6 Sita Venkateswar, *Development and Ethnocide: Colonial Practices in the Andamans* (Copenhagen, 2004).

7 See for example, Ian Tattersall, author of *Extinct Humans* and curator of physical anthropology at the American Museum of Natural History, New York, who describes the Andamanese as 'a people who never stood a chance' (cited on the back cover of Madhusree Mukerjee, *The Land of Naked People: Encounters with Stone Age Islanders* (London, 2003).

8 Satadru Sen, in his *Savagery and Colonialism in the Indian Ocean: Power, Pleasure and the Andaman Islanders* (London, 2010), moves some way to redressing this lack of periodization, although he focuses mainly on the changing approaches of the British *to* the Andamanese.

9 Douglas Haynes and Gyan Prakash, 'Introduction: The Entanglement of Power and Resistance', in Douglas Haynes and Gyan Prakash (eds), *Contesting Power: Resistance and Everyday Social Relations in South Asia* (New Delhi, 1991), p. 12.

10 See Claire Wintle, 'Objects, Images, Imaginings: New Perspectives on the Material and Visual Culture of the Andaman and Nicobar Islands', *Journal of Museum Ethnography*, 20 (2008), p. 151.

11 Ranajit Guha, *Elementary Aspects of Peasant Insurgency in Colonial India* (New Delhi, 1983), p. 15; Gayatri Chakravorty Spivak, *In Other Worlds: Essays in Cultural Politics* (New York, [1997] 2006), p. 287.

12 For example, Peter van Dommelen, 'Colonial Matters: Material Culture and Postcolonial Theory in Colonial Situations', in Chris Tilley (ed.), *Handbook of Material Culture* (London, 2006), pp. 112–13.

13 Clare Anderson, *The Indian Uprising of 1857–8: Prisons, Prisoners and Rebellion* (London, 2007), pp. 129–31.

14 R.H. Colebrooke, cited in Maurice V. Portman, *A History of Our Relations with the Andamanese: Compiled from Histories and Travels, and from the Records of the Government of India*, vol. I (Calcutta, 1899), p. 69.

15 R.C. Tytler, Superintendent of Port Blair, 1862–1864, cited in Portman, *A History of Our Relations with the Andamanese*, vol. I, p. 381.

16 'Report on the Andamanese Department for July, 1895', Proceedings Port Blair, 1895, P/4725, p. 56; 'Report on the Andaman Home for June, 1887', Proceedings Port Blair, 1887, P/2898, p. 50.

17 Sen, *Savagery and Colonialism in the Indian Ocean*, Introduction.

18 Edward Horace Man instigated the adoption of Western clothes in 1876 (Maurice V. Portman, *A History of Our Relations with the Andamanese: Compiled from Histories and Travels, and from the*

Records of the Government of India, vol. II (Calcutta, 1899), p. 606); the teaching of English was instigated under Henry Corbyn in 1863, but had been condemned by Portman by 1899 (ibid., pp. 377–8). The reliable, consistent supply of food was apparently one of the most enduring reasons for Andamanese attendance at the homes. The use of grain as a supplement to their regular diet is particularly referenced as having occurred under J.N. Homfray's care between 1864 and 1874 (B. Ford, cited in Portman, *A History of Our Relations*, vol. II, p. 540) but was probably a constant throughout the British period.

19 Cited in Portman, *A History of Our Relations*, vol. II, p. 564. See Sen, *Savagery and Colonialism in the Indian Ocean*, ch. 4, for an analysis of how the relationship between work and the rhetoric of civilization shifted subtly during the second half of the nineteenth century. Eventually, it was considered that the Andamanese could not be 'civilized' and therefore hard labour, especially agriculture, was stopped. The production of indigenous manufactures for sale continued throughout this time, however.

20 'Report on the Andamanese Department for January, 1875', Proceedings of the Revenue and Agricultural Department, Asia, Pacific and Africa Collections, British Library, 1875, P/528, p. 27; Portman, *A History of Our Relations*, vol. II, p. 601; Violet Talbot Clifton, *Pilgrim to the Isles of Penance: Orchid Gathering in the East* (London, 1911), p. 86.

21 W.H. Burt, untitled typescript detailing life on the Andaman Islands, undated [1914], Department of Africa, Oceania and the Americas' Archive (Eth Doc 1714), British Museum; Frederick Taylor, 'Native Life in the Andaman Islands: A British Penal Settlement for India, with Pictures from Photographs Taken for the Writer', *Century Illustrated Monthly Magazine*, 82/6 (1911), pp. 897–8.

22 'Andaman Home Report for August 1878', Proceedings of the Superintendent of Port Blair and the Nicobars, Asia, Pacific and Africa Collections, British Library, 1878, P/1144, p. 37; Portman, *A History of Our Relations*, vol. II, p. 700.

23 See David Tomas, 'An Ethnography of the Eye: Authority, Observation, and Photography in the Context of British Anthropology 1839–1990' (unpublished doctoral thesis, Department of Anthropology, McGill University, 1987), pp. 299–300; Venkateswar, *Development and Ethnocide*, p. 102, and Satadru Sen, *Disciplining Punishment: Colonialism and Convict Society in the Andaman Islands* (Oxford, 2000), pp. 232–4.

24 Portman, *A History of Our Relations*, vol. II, p. 597.

25 Richard Carnac Temple, *Remarks on the Andaman Islanders and Their Country* (Bombay, 1930), p. 62; A.R. Radcliffe-Brown, *The Andaman Islanders* (New York, [1922] 1964), p. 19.

26 Portman, *A History of Our Relations*, vol. II, p. 700.

27 Vishvajit Pandya, 'Deforesting Amongst Andamanese Children: Political Economy and History of Schooling', in Barry S. Hewlett and Michael E. Lamb (eds), *Hunter-Gatherer Childhoods: Evolutionary, Developmental, and Cultural Perspectives* (New Brunswick, NJ, 2005), pp. 395–6.

28 Sen, *Disciplining Punishment*, p. 157.

29 Tomas, 'An Ethnography of the Eye', p. 299.

30 Ruth Phillips, *Trading Identities: The Souvenir in Native North American Art from the Northeast, 1700–1900* (Seattle, WA and Montreal, 1998), p. 17. In turn, Phillips draws on Mary Louise Pratt, *Imperial Eyes: Travel Writing and Transculturation* (London, 1992).

31 These details are drawn from cross-referencing descriptions in Radcliffe-Brown, *The Andaman Islanders*, pp. 476–80 and Edward Horace Man, 'On the Aboriginal Inhabitants of the Andaman Islands (Part III)', *Journal of the Anthropological Institute*, 12 (1883), pp. 330, 400 and 406.

32 See Vishvajit Pandya, *In the Forest: Visual and Material Worlds of Andamanese History (1858–2006)* (Lanham, MD, 2009), pp. 113–15 and 124–5; and Vishvajit Pandya, *Above the Forest: A Study of Andamanese Ethnoanemology, Cosmology, and the Power of Ritual* (New Delhi, 1993), pp. 123–5 and 227–9, for a detailed examination of how the Ongee of Little Andaman continue to use both clay and shells to instigate communications with the spirit world and initiate particular events integral to their life cycle.

33 Radcliffe-Brown, *The Andaman Islanders*, p. 44. This is not the case for all adornments, such those incorporating human bone (Vishvajit Pandya (Dhirubhai Ambani Institute of Information & Communication Technology), in an interview with the author, 26 September 2008).

34 See, for example, those published in Edward Horace Man, *A Dictionary of the South Andaman (Aka-Bea) Language* (Bombay, 1923), reprinted from the *Indian Antiquary* and those reproduced in Elizabeth Edwards, 'Science Visualized: E.H. Man in the Andaman Islands', in Elizabeth Edwards (ed.), *Anthropology and Photography 1860–1920* (London, 1992), e.g. pp. 110, 111 and 112.

35 W.H. Burt, untitled typescript.

36 Notably, the image includes the highly unlikely case of a woman 'shooting' a bow and arrow. See Sen, *Savagery and Colonialism in the Indian Ocean*, pp. 12–13 on the performance of 'savagery' in the Andaman Islands.

37 But not by Taylor himself, who describes the Andamanese adorned in this way when he arrived at the home at Haddo unannounced (Taylor, 'Native Life in the Andaman Islands', pp. 896–7).

38 This is supported by closer comparative analysis of Man's earliest collections in the founding collection of the Pitt Rivers Museum and Taylor's collections at the American Museum of Natural History.

39 See, for example, a ceremonial waistband with blue and turquoise beads from Great Andaman collected by Reverend W.R. Park in 1914 at the British Museum (1963.AS4.12).

40 Chris Gosden and Chantal Knowles, *Collecting Colonialism: Material Culture and Colonial Change* (London, 2001), p. 22.

41 Radcliffe-Brown, *The Andaman Islanders*, pp. 42–3.

42 Pandya, *Above the Forest*, p. 276.

43 Edward Horace Man, *On the Aboriginal Inhabitants of the Andaman Islands with a Report into the Languages of the South Andaman Island by A.J. Ellis* (London, 1932), p. 216.

44 For example, Allen Thomson, 'Description of Andamanese Bone Necklaces', *Journal of the Anthropological Institute of Great Britain and Ireland*, 11 (1881), p. 304.

45 In 1899, it was noted that arrows were produced and kept in the 'Home Godowns for sale as curiosities'; 'Report on the Andamanese Department for January 1899', Proceedings of the Superintendent of Port Blair and the Nicobars, Asia, Pacific and Africa Collections, British Library, 1899, P/5616, p. 171.

46 On Nelson Graburn's scale of the 'arts of acculturation', this stage of production is termed 'commercial fine arts', where objects are for sale, but still culturally embedded (Nelson Graburn, 'Introduction: The Arts of the Fourth World', in Nelson Graburn (ed.), *Ethnic and Tourist Arts: Cultural Expressions from the Fourth World* (Berkeley, CA, 1976), pp. 5–7.

47 David Gauntlett, *Creative Explorations: New Approaches to Identities and Audiences* (London, 2007), p. 19. Here, I draw on Gauntlett's broad definition of creativity which involves the 'physical making of something, leading to some form of communication, expression or revelation' (ibid., p. 25).

48 For example, David Gauntlett, *Making Is Connecting: The Social Meaning of Creativity, from DIY and Knitting to Youtube and Web 2.0.* (London, 2011); Andrew Harrison, *Making and Thinking: A Study of Intelligent Activities* (Brighton, 1978).

49 Pandya, *Above the Forest*, p. 130; Radcliffe-Brown, *The Andaman Islanders*, p. 323.

50 Radcliffe-Brown, *The Andaman Islanders*, p. 195.

51 Pandya, *Above the Forest*, p. 255.

52 Richard Sennett, *The Craftsman* (London: 2008), p. 8; Christopher Tilley, 'Introduction', in Christopher Tilley et al. (eds), *Handbook of Material Culture* (London, 2006), p. 61.

53 Cited in Portman, *A History of Our Relations*, vol. II, p. 597.

54 Valentine Ball, 'Visit to the Andamanese "Home," Port Blair, Andaman Islands', *Indian Antiquary*, 3 (1874), p. 171.

55 See Radcliffe-Brown, *The Andaman Islanders*, p. 47 for details of the role women played in community relations.

56 Ball, 'Visit to the Andamanese "Home"', pp. 171–2.

57 For example, Gauntlett, *Making is Connecting*, p. 2.

58 Radcliffe-Brown, *The Andaman Islanders*, p. 76.

59 See observations made in ibid., pp. 41 and 50.

60 Tomas, 'An Ethnography of the Eye', p. 299.

61 See, for example, Dan McDougall, 'How a "Stone Age" Tribe is Keeping the Modern World at Bay', *The Week*, 18 February (2006), pp. 44–5, and Mukerjee, *The Land of Naked People: Encounters with Stone Age Islanders*.

62 Radcliffe-Brown, *The Andaman Islanders*, p. 122.

63 See Pandya, *In the Forest*, pp. 117 and 151–99.

64 Clare Anderson, 'Colonization, Kidnap and Confinement in the Andamans Penal Colony, 1771–1864', *Journal of Historical Geography*, 37 (2011), pp. 76 and 79, and Satadru Sen, 'Policing the Savage: Segregation, Labor and State Medicine in the Andamans', *Journal of Asian Studies*, 58/3 (1999), pp. 753–73.

65 See a discussion of the concept of agency in Chris Barker, *The Sage Dictionary of Cultural Studies* (London, 2004), pp. 4–5.

66 Glenn Adamson, 'Craft in Action: Life, Art, Design: Section Introduction', in Glenn Adamson (ed.), *The Craft Reader* (London, 2010), p. 459.

67 Notably, even in Pratap Dutta's heartbreaking descriptions of the Great Andamanese in the 1970s, when they were in their most fragile state to date, he highlights how they still preferred to live by traditional modes of hunting and collecting (Pratap C. Dutta, *The Great Andamanese: Past and Present* (Calcutta, 1978), p. 40).

8

Crafting Inclusion for 'Invalid' Women: The Girls' Friendly Society Central Needlework Depôt, 1899–1947

Vivienne Richmond

> This work has been going on continuously for forty-seven years … long before the need for occupational therapy was generally recognised. It has been pioneer work, and there is still no similar organisation for disabled women.[1]

The elite magazine *The Lady* was celebrating, in 1946, the work of the Girls' Friendly Society Central Needlework Depôt (the Depôt), established in 1899 to provide single, poor, blind and 'invalid' women with paid home training and employment in craftwork. Home-working schemes for the disabled had existed in Britain since at least the mid-nineteenth century, but most were restrictive, assuming the workers capable of only the most basic tasks.[2] The Depôt, in contrast, sought to provide employees with skilled, purposeful employment and, as an all-female employment scheme for disabled women, within an all-female mainstream social organization, the Depôt was a unique enterprise.[3] Its double origins lay in the social purity and craft revival movements of the late nineteenth and early twentieth centuries and the critical history of the Depôt uncovers new aspects of the work of both movements.

Little has been written about the history of disabled women. Much of the existing historiography of disability leads teleologically to the injuries sustained by men in the First World War and the medical, surgical and psychiatric developments as well as the new social awareness and conceptualization of disability that resulted.[4] Women tend to enter the picture as the subsequent beneficiaries, or recipients, of these developments. This chapter, which draws on the extant records of the Depôt, begins with a summary of its history and connections to the craft revival movement.[5] It continues by considering the position of unmarried disabled women during a period when the prevailing ideology deemed marriage and motherhood with support by a male breadwinner the natural and desirable destiny of all women. Finally this chapter examines the relationship between the Depôt and its workers to reveal an organization simultaneously paternalistic and emancipatory that, through skilled craftwork, diminished the social isolation and increased the autonomy of a group marginalized not only through disability, but because they were poor, female and single.

The Central Needlework Depôt

The Girls' Friendly Society (GFS) was an Anglican social purity organization. Founded in England in 1875 to promote pre-marital chastity among working-class girls, it was managed, centrally and locally, by middle- and upper-class women, termed Associates. The Society's main target was the large number of young women who left rural homes for work in the cities as domestic servants, the occupation of one in four of the employed female population at mid-century.[6] Fearful that away from the protection of family and community these girls, frequently only 12 or 13 years old, were often lonely, isolated and, therefore, vulnerable to sexual predators, the Society offered friendship, guidance, wholesome entertainments and education through club rooms and branch meetings, travel escorts, convalescent and holiday homes, residential lodges and employment bureaux.[7] Since the assumed loss of virginity on marriage terminated GFS membership, most members were young, and at the Society's peak in 1913 they numbered nearly 200,000 in England and Wales.[8] Among these, domestic servants formed the largest occupational group, but the GFS welcomed all chaste working-class females, whether in employment or not, and operated a Department for Sick Members with short- or long-term illnesses and disabilities. It also had an Industrial Department, which trained Members in 'useful, interesting, and remunerative handicrafts', and 'invalids and blind Members so that they could to a certain extent support themselves'.[9]

From 1885 the Industrial Department began to hold an annual sale of needlework produced by disabled members, many of whom were otherwise excluded from paid work.[10] Four years later it opened a Home of Industry for them in Gloucestershire, which also served as a depôt for the sale of work made by disabled members at home.[11] In 1899 the latter function transferred to the newly established Central Industrial Depôt (subsequently renamed the Central Needlework Depôt), which operated from a room at the Society's London headquarters. Within months its work 'had so increased it was absolutely necessary to hire a second room and get an extra assistant' to aid the manageress.[12] By the 1920s, the Depôt employed 500 'Wonderful Workers', as well as art teachers, a publicity secretary and an accountant, and it enjoyed royal patronage. When the Society moved into purpose-built headquarters in Victoria in 1925, the Depôt occupied much of the first floor where it had a showroom equipped 'for the display of the various kinds of garments, fancy articles and baskets' and a series of rooms for cutting out, packing and staff offices.[13]

Despite its seeming success, the Depôt was never financially independent. In 1907, the Society's annual report explained that successful sales and the addition of 72 new workers that year notwithstanding, the Depôt would always require support from subscriptions due to the costs of teaching new recruits and preparing the work before despatch to the employees.[14] By 1936 the GFS itself was in financial trouble and the

Depôt was soliciting funds through a radio appeal. The introduction of workers' Health Insurance, for which the Society took responsibility, added to the financial burden. The war, clothes rationing, the imposition of Purchase Tax, a drop in subscriptions, trade and wage regulation, as well as the difficulty of competitively pricing hand-made goods in an age of mechanized mass-production, added to the problems.[15] By the end of 1945 the Depôt was verging on insolvency and the 1946 *Lady* article was clearly an unsuccessful last-ditch attempt to drum up support.[16] In 1947, the year before the introduction of the welfare state, the Depôt closed, having employed over 1,000 workers during the course of its existence.[17]

The Depôt and Craft Revival

In 1876, the year after the GFS was founded, C.J. Leland published *The Minor Arts*, in which he advocated village classes for craft instruction. From this issued the Cottage Arts Association, which, in 1884, became the Home Arts and Industries Association (HAIA). Through the home manufacture of craftwork for sale, the HAIA sought to alleviate poverty, bring 'civilization' and beauty to the lives of the poor, and challenge the growing supremacy of undifferentiated industrially produced goods.[18] It was preceded, in 1883, by the Donegal Industrial Fund, examined by Janice Helland in Chapter 6 of this volume, and soon followed by Irish, Scottish and Welsh home-industry associations.[19]

The development of the Depôt was closely bound up with, and shared many of the characteristics of, this craft revival movement in which the workers' 'overriding designator', says Helland, 'was undoubtedly class and gender (peasant and poor, female with no status)'.[20] The Depôt workers were not peasants, but being working-class, poor, female and disabled, they were certainly lacking in status. According to Helland, the home industries were also:

> concerned for the morality of the workers, and all owed the revived interest in their projects to the promotion of exhibitions of women's work, increased accessibility of women's magazines, the growing interest in arts and crafts, and to the patronage of well-educated urban or wealthy landed women.[21]

The GFS was similarly dependent on the support of elite women. Of the 41 Associates elected to its central or regional presidential positions in 1901, 15 were titled and these women, moving in the same circles as those who orchestrated and patronized the craft revival movement, would have been aware of the various ventures being established.[22]

While the HAIA favoured the sale of goods at public exhibitions, the Irish and Scottish regional associations preferred private sales in aristocratic homes. The GFS adopted both techniques to bring 'before the public the really beautiful and useful work' the Depôt could supply.[23] Its initial 1885 sale was held in May, three months after the Bristol Exhibition of Women's Industries, and two months before the first HAIA sale.[24] The venue was the Knightsbridge residence of Lady Egerton of Tatton, the GFS Diocesan President for Chester, and even after the Depôt acquired the magnificent purpose-built showroom in the 1920s it still held sales 'in the private houses of friends'.[25]

But it also sold at public festivals and exhibitions, and, from at least 1900, the Depôt had a 'stall of Invalid and Blind Members' work' at the HAIA annual Exhibition at the Albert Hall, cementing its connection with the craft revival movement.[26] This drew the members into a wider community of craftworkers and craftwork initiatives, the chief concern of which was the alleviation of poverty, not bodily impairment or difference. And just as the home industries brought training and employment to geographically isolated women, so by drawing the Depôt within its embrace the HAIA integrated women isolated by disability.[27]

One of the Depôt's stated aims was 'to secure a better sale for the work by bringing it more prominently before the public'.[28] Helland notes that the craft revival movement was facilitated by 'a growing awareness of the usefulness of exhibition and display' and a concern to 'promote and sell objects made by the poor', who lacked 'access to metropolitan venues and wealthy customers'.[29] This was achieved not only by selling at large exhibitions, but also by the home industries' establishment of their own shops and showrooms in the fashionable West End of London. The Donegal Industrial Fund, for example, opened a shop in New Cavendish Street in 1883, which transferred to Wigmore Street two years later and, at the same time, held a month-long exhibition in a Regent Street art gallery. Just a year before the creation of the Depôt, the Welsh Industries Association opened its Sloane Square showroom.[30] All were just a mile or two away from the GFS's Victoria headquarters, providing professional models for its own display and sales techniques.

Helland also stresses the regional home industries associations' prioritization of fashion, in which respect they differed from the HAIA.[31] The Depôt fell somewhere between the two. An illustration of the Depôt's showroom suggests a rather homely selection of goods, but the accompanying text noted 'the importance of "watching the market" … noting the changes in fashion and fitting the supply to meet the demand'.[32] The Depôt made wedding trousseaux something of a specialization, supplying part of Princess Mary's when she married in 1922, and it was a regular supplier of her nightgowns and camisoles. An extensive correspondence between the Depôt and the Palace, in which samples of garments and lace were sent for approval, colours discussed, and small size alterations made, as well as the quantities ordered – a dozen at a time – all indicate that these were garments the young princess wore, not mere token orders.[33] Royal support of the Irish lace industries, says Anthea Callen, 'set fashions among the wealthy classes'.[34] It is rather more doubtful that Princess Mary's patronage of the Depôt set the fashion for élite underclothing, but clearly it was hoped that her support would boost sales.

Unmarried Women, Work and Disability

The GFS saw marriage and motherhood as women's ultimate goal, but accepted that this was not the destiny of all. Although it did not support female suffrage, it did, as Brian Harrison maintains, contribute to female emancipation through 'expanding women's self-respect, and in pioneering new opportunities for careers and usefulness' for women of all classes. And so, while the moral regulation of 'social purity' was the Society's foundation, it also encouraged education and personal and professional self-development among the members. Anne Anderson detects a parallel political conservatism among the female 'cultural philanthropists' of the home industries associations, but says that through their involvement with these organizations, art 'opened up new horizons in voluntary work' for unmarried or childless élite women. Harrison similarly notes that the GFS was especially adept at finding meaningful occupation for spinsters.[35] Like Anderson, Harrison was referring principally to middle- and upper-class women – those who ran the GFS – but the same was true for the members. The Depôt archive contains 68 application forms from prospective workers in England and Wales. Their ages, presented in Figure 8.1, show that the majority were adults and their GFS membership meant they were necessarily spinsters.

Age of applicant	15–19	20–29	30–39	40–49	50–59	61	Not stated
Number of application forms (total 68)	7	15	17	13	6	1	9

8.1 Ages of prospective workers stated on 68 extant GFS Central Needlework Depôt application forms[36]

In the late nineteenth and early twentieth centuries spinsters were, at best, pitied, and at worst, scorned and ridiculed, and the disabled spinster still more so. Susan M. Schweik cites an American judge who, in 1910, deemed it worse for a woman to be a 'cripple' than a man. Her assumed loss of 'physical attractiveness' reduced her marriage chances and her abhorrence at exhibiting her 'infirmity' made it difficult to earn the necessary money for self-support. '[T]he crippled woman who did not shrink from exhibiting herself', says Schweik, 'could hardly be recognized as a woman'.[37] Cindy LaCom found that when disabled bodies begin to appear in fiction, in the nineteenth century, they are mostly female and linked to fears about sexuality and the maternal. Women who were 'incapable of overseeing the domestic sphere and bearing children', she says, 'faced pity, suspicion, and/or hostility', and 'physical disabilities that "disfigured" women and kept them from becoming mothers were literally and symbolically crippling'.[38] The female invalid was an in-valid woman.

Furthermore, in the late nineteenth and early twentieth centuries, eugenist fears about racial degeneration, which commonly coupled physical incapacity with mental impairment, poverty, criminality and other anti-social behaviour, sought to prevent disabled people marrying and reproducing.[39] Anne Borsay highlights criticism of interpretive models of disability history that omit 'structural factors besides impairment – social class, gender, age and ethnicity – the interplay of which also shapes experiences of exclusion'.[40] This is precisely pertinent to the Depôt workers, who were marginalized not only by their disability, but because they were poor, female and unmarried.

However, what, for the Depôt's purposes, constituted an 'invalid' woman is unclear. Each prospective worker had to submit a sample of her work with an application form, endorsed by her Associate, stating how long she had been a member of the GFS, her age, the type of work she wished to do and the nature of her 'illness'. In the late nineteenth and early twentieth centuries, except for the blind, deaf and dumb, people with restricted physical capacity were termed sick, invalid or crippled, while those with a learning difficulty were identified as defectives or imbeciles, and discussion of disability is complicated by the lack of a precise, agreed definition. The 68 Depôt applications were submitted between 1898 and 1946 and cite a wide range of illnesses and disabilities, all physical or 'nervous'. While some are clear, such as blindness, rheumatism and amputations, others are less so: 'Nervous trouble (Diagnosed as nervous indigestion)'; 'Is Anaemic & delicate generally'; 'Doctors cannot decide internal weakness'; 'Strength out of both legs & internal trouble'. Some can be linked to illnesses such as polio and tuberculosis, then common in Britain: 'Infantile paralysis (leg in irons)'; 'spinal caries'.[41]

The 1901 census returned 17,899 blind, deaf and dumb females aged ten and above, of which only just under 3,000 were 'engaged in occupations', although some 10,000 were unmarried.[42] Rosemarie Garland Thomson proposes that 'the most enduring form of segregation [for disabled people] has been economic', and Steve Humphries and Pamela Gordon emphasize the very high levels of unemployment among them in the first half of the twentieth century. This resulted in 'absolute poverty for many', and the clustering of people with disabilities in the poorest areas of large cities where, of necessity, they might turn to begging or crime.[43] As such, official definitions of physical and mental incapacity were primarily concerned with an individual's capacity for employment and financial self-reliance. Legislation in the 1890s to provide education for blind, deaf, epileptic and 'defective' children aimed 'to prevent them becoming not only "a burden to themselves but a weighty burden to the state"'.[44]

Similarly the apparently unique 1911 census of cripples in Birmingham, England, defined a cripple as 'A person whose (muscular) movements are so far restricted by accident or disease as to affect his capacity for self-support'. Its analysis of the 1,001 cripples aged 16 and over in Birmingham categorized them according to the cause of their disability and their ability to work 'under ordinary conditions', at 'a central workshop', at home, or not at all, the last comprising 53 per cent. Further analysis considered their capacity for self-support, while the 1,006 Birmingham cripples under 16 years old were divided into those who were and were not working, and the likely future capacity of the latter for paid employment.[45] Borsay documents that in 1920 a Board of Education spokesman justified the expense of children's orthopaedic hospitals on the basis that with 'proper treatment' three-quarters of the country's 'child cripples' could be turned into 'satisfactory workers' and 'self-supporting citizens' rather than 'a burden on their relatives and the community'.[46]

Inevitably, therefore, provision for the disabled focused on finding them work. In her 1914 study of the care of crippled children in America, Edith Reeves surveyed their occupational training in the various institutions they attended. She was concerned not only with the suitability of the employment to the individual's capacity, but also its profitability – would it enable self-support? Reeves insisted that the children should be taught skilled occupations, such as clerical work, in preference to unskilled low-paid trades where they would have difficulty competing with the speedier output of able-bodied workers. But she recorded the prevalence of training in craftwork, the challenge of making a living from it, and the 'practically unanimous' opinion 'that needle trades' solved the employment problem for disabled girls, and the same was true in Britain.[47]

The Depôt provided members with training, mainly via correspondence courses, in a variety of crafts including crochet, leather and basketwork, but sewing and knitting predominated.[48] Yet it would be a mistake to dismiss the Depôt as simply another scheme that condemned disabled females to low-skilled work. Needlework was a compulsory subject for all working-class girls in state-funded elementary schools. Indeed, more time was devoted to it in the girls' curriculum than to any other single subject, but instruction was restricted to utilitarian plain needlework, with embroidery and 'fancy work' – the privilege of elite women – expressly forbidden.[49] The schooling of girls with disabilities was likely to be interrupted through illness and visits to doctors and clinics, but the Depôt not only made up for deficiencies in the members' previous education, and paid them while they trained, but encouraged them to undertake decorative work.[50] As a history of the GFS, written to celebrate its golden jubilee in 1925 explains, one of the Depôt's aims was to 'raise the standard of work by supplying first-rate patterns and skilled instruction for the workers'. The result, claimed the author, was that the 'girls were taught new industries, and the market thus relieved of the plethora of knitted goods, in so many cases the sole resort of the invalid'.[51] The exquisitely hand-stitched sample of work attached to Ellen Brewer's 1898 application demonstrates the high standard of work of which some workers were capable. Also, the inextricable connection between needlework and femininity in the Victorian and Edwardian periods meant that its performance helped to alleviate the desexualizing effect of the workers' spinsterhood.

Borsay argues that social inclusion for disabled people 'was conditional upon fulfilling the requirements of economic rationality and participating effectively in the labour market', but that 'In the nineteenth century, charitable services perpetuated this disadvantage by promoting divisive segregated training and sheltered workshops that traded in outdated, manual skills and associated impairment with poorly paid, low-status work'.[52] While the archive retains little substantive evidence of the workers' wages, references to small earnings, the Depôt's stated wish to supply the market with 'really good work at moderate prices' and its lack of independent financial viability all point to low pay.[53] Even by the time it closed in 1947, only 15 of the Depôt's 61 workers were earning more than £10 a year.[54]

However, a comprehensive evaluation of the Depôt workers' economic and occupational segregation requires comparison with able-bodied working women. The 1901 census returned 340,582 females aged ten and above working as dressmakers. The only employment with a greater number of female workers was domestic service, with almost 1.3 million, and the next largest after dressmakers was laundry and washing with just under 200,000.[55] Humphries and Gordon say that in Britain, until the Second World War, disabled girls 'were generally taught dressmaking, embroidery, laundry work or domestic service', in which case, at least in the early twentieth century, their training mirrored the prevalent occupations of their able-bodied counterparts.[56] Furthermore, 82 per cent of the dressmakers in the 1901 census were unmarried or widowed, and

43 per cent of these worked at home.[57] Home employment in needlework was, then, scarcely a marginal occupation for single women.

Nineteenth-century needleworkers were notoriously badly paid, and, as Clare Rose says, 'exploitative practices were endemic to the nineteenth-century clothing trade, with scandals over death through overwork recurring in 1863 and in 1904'.[58] In this respect, Depôt employees had a slight advantage over their able-bodied counterparts since the Depôt took account of the limitations their health might place on their productive capacity. Delayed work was met either with assurances that there was no hurry; or, in the case of special orders, a request to return it and a promise of replacement with a stock item for completion at the worker's convenience. Pay was for piecework, the characteristic arrangement of the craft revival movement, says Callen, so those unable to work were unable to earn.[59] But workers who returned items to the Depôt unfinished or badly executed were paid for what they had done, even if the item was unsaleable. In the 1920s, Violet Hill, for example, repeatedly returned substandard work and was reprimanded because the manageress believed it resulted from carelessness, rather than excusable ill health. But she continued to be supplied with work and paid for it, even though it had to be made good by another employee.[60]

To say that the Depôt workers were, like others in the same industry, poorly paid does not negate Borsay's assertion that disabled people were largely confined to low-paid work or excuse the practice. Nor does the parallel between the occupations of disabled and able-bodied women in the early years of the Depôt's existence ignore the increasingly anachronistic focus on craftwork for disabled women and their exclusion from the expanding female occupations, such as the clerical work advocated by Edith Reeves. But it does emphasize that poor pay was a scandalous fate the Depôt workers shared with a large number of able-bodied working-class females and suggests that, on this issue, the determining factor was not overt discrimination but their trade and the Depôt's limited resources. This is supported by minutes of management meetings that record concern about the workers' low wages and improvements where possible. In 1944, for example, the 'invalids were specially helped by a big increase in the amounts paid to them for work done'.[61]

The Depôt and the Workers

Elizabeth Cumming, while noting the connection between the craft revival and the Arts and Crafts movement, asserts that the former 'had no sympathy for the rallying call of [William] Morris's socialism', and the GFS was careful to maintain a clear separation between classes even as it fostered relationships between them.[62] A 1928 promotional booklet asked readers to:

> Think of the monotony of a sick bed, the weary sense of helplessness, of dependence upon others. Then picture the delight of being trained to some new handicraft, or perfected in an old one, the awakening to new interests, the sheer joy of handling delicate, daintily-coloured fabrics, the pride and pleasure with which the kind words of approval from the Manageress are received … Surely the joy of self help must count for much in a life of suffering.[63]

In her study of the Dun Emer Guild, a branch of the Irish craft revival, Elaine Cheasley Paterson notes that its products 'were beyond the means of most of the working-class women who produced them'.[64] Similarly, the Depôt employees' assumed pleasure in working with luxurious fabrics doubtless lay in the contrast with the inferior materials of their own clothes. Their perceived lack of independence, anticipated gratification at the Manageress's 'approval', presumed 'suffering' and general gratitude for all that is being done for them displays the Society's paternalism and accords with Schweik's assertion that charities' accounts of their work focused 'on the good social worker, the wise and superior friend', rather than on the actions and agency of the disabled subjects.[65]

Susan Mumm questions the frequent claim that 'charitable work infantilized and subordinated its recipients', but this is a charge easily laid against the Depôt.[66] It constantly referred to the adult workers as 'girls', chastized them in language more usually reserved for children when they submitted work perceived to be careless, and reminded them of their reliance on the Depôt. 'Dear Violet', wrote the manageress in 1929 to the 40-year-old Miss Hill:

> I was very disappointed indeed on going into the workroom half an hour ago to see the work which you have sent in to-day ... It is very naughty of you to send in work hurriedly as you know that we always do our best to give you as many orders as possible as we know that you need the money.[67]

While the Depôt accepted applicants' own claims of disability, albeit endorsed by their Associates, rather than requiring a professional medical diagnosis, there was otherwise a lack of worker agency concerning information about their health. When applicants had previously had a communicable disease, such as tuberculosis, the Depôt required certification that they were free from infection, but this was usually requested from her Associate or doctor rather than from the prospective employee herself.[68] One Associate wrote confidentially to the Depôt Manageress, again in 1929, about Elsie Bishopp, a 39-year-old worker whose leg had been amputated apparently because of a tubercular knee. The Associate had heard, though not from Bishopp, that there was 'some questioning of a lung' and Bishopp was to be X-rayed. She acknowledged that this might have been 'gossip', but anyway suggested requesting a certificate to 'make sure'. In the subsequent exchange of correspondence it transpires that, between them, the Associate and the Manageress discussed Bishopp's health with her doctor, a hospital ward sister and 'the T.B. Aftercare people'. The latter eventually confirmed that Bishopp's doctor had signed a statement saying that her amputation resulted from a knee injury and while multiple operations had reduced her to 'a condition of chronic nervous invalidism', there was 'No conclusive evidence of tuberculosis at any time'.[69] Not only had Bishopp's health, on the basis of hearsay, been the subject of discussion and investigation among this varied group of people, but it had all been in pursuit of confirmation that she was now free from a disease she had probably never contracted.

Bishopp's views on the matter, if she was aware of it, are not recorded, but Mumm also argues that recipients of charity retained the power to refuse aid. She questions therefore whether 'acceptance of assistance in the past was always such a thoroughly demeaning experience'.[70] The validity of her argument is tempered by the paucity of real alternatives, but it is nevertheless clear that lack of choice and (limited) financial gain were neither the only nor necessarily the main reasons why the Depôt workers continued their employment. They valued the work also for providing purpose and a distraction from troubles. Elsie Bishopp wrote frequently to the Manageress about her troublesome stump and when, in 1933, she learned that, disappointingly, she would have 'to leave the artificial leg off – and go back on crutches', she began by saying that the work had 'been such a help in ever so many ways' over her recent 'worrying time as regards the stump etc'. A few years later, learning that a deterioration of her health would not prevent her from working, she was 'so glad, as I love it all. And after all this time I shouldn't like having to sit twiddling my thumbs'. Similarly, Lizzie Brabban, who had worked for the Depôt for 30 years, wrote to the Manageress:

> I was very ill for a time & had to have complete rest ... Now that I am keeping better I sometimes wish I had some work to do as I miss it very much. I was always so fond of it, so if you could send me a little pleasant order I could manage it.[71]

In a 1992 oral history documentary of disability in the first half of the twentieth century, feelings of isolation – in school, socially and in the workplace – even when surrounded by people, surfaced repeatedly.[72] For some Depôt workers, correspondence with the Manageress was clearly also a source of companionship. Elsie Bishopp, for example, sent an update on her progress with some work and on her health, but continued with news 'of the arrival of a new niece. My youngest sister's first, and her name is "Anne". I'm ever so pleased – such a nice baby she must be 8lbs'. As in Bishopp's case, the social information was usually an addendum to a business letter, but Bella Bateman, for example, wrote to the Manageress in 1933 solely to seek solace during the 'terrible & great loss that we are passing through', the unexpected death of her mother. She would, she said, 'love a line from you dear if you could write to me, to help to comfort us'.[73] It therefore appears that at least some of the workers were unconscious of, or untroubled by, the Depôt's paternalism, considering the Manageress as much a friend and counsellor as a supervisor, and valuing the work not simply for the income it brought, but for its intrinsic interest and occupation.

Conclusion

The Lady correctly identified the Depôt's work as pioneering. Women, says Borsay, were 'marginal to paid work because "most policy makers assumed that all females, regardless of their physical status, ought to be dependent on fathers, brothers, or husbands"'.[74] The GFS did not make that assumption; instead it created the Central Needlework Depôt to offer unmarried disabled women training and employment. Where other schemes focused on what they believed disabled people could not do, the Depôt sought to identify and develop what they could. And in our haste to justly condemn ignorant policies that restricted disabled people to low-skilled manual occupations, we must not forget that craftwork could also be highly skilled, and a source of pleasure and pride to its producers. The offence lay in its abysmal remuneration.

While our twenty-first-century sensibility might balk at the Depôt's paternalistic and often patronizing approach, its connection with two mainstream organizations for working-class females, the Girls' Friendly Society and the home industries movement, made craftwork the vehicle for a degree of social and occupational inclusion for a group of women who were highly marginalized, through poverty, spinsterhood and disability. Anderson and Cumming both note the commitment of upper- and middle-class supporters of the craft revival movement to 'workforce welfare'.[75] This was also a priority for the GFS, which had a sincere wish to provide practical support for disabled members. In addition to the Depôt, it provided many with grants and pensions and from 1891 ran a postal Braille library. Individual branches 'adopted' invalid members, sending them letters, books, flowers, food and equipment, as well as visiting where possible. And while these activities may have been orchestrated by the élite women who ran the organization, they were performed by the working-class members, creating peer contact between them and the 'invalids'.

Cheasley Paterson says of the Dun Emer Guild that it challenged 'restrictive social codes' by positioning the peasant women it employed 'within the public realm of industry and paid work'.[76] The Depôt did the same for its workers. Although goods were advertised as the produce of disabled women, this was not to promote sympathy purchases of substandard work, but to emphasize its quality and the workers' abilities. At the 1900 Home Arts Exhibition at the Albert Hall, work from the Invalid and Blind Members' stall won 71 awards.[77] By 1907 this had increased to 159, 'including the gold cross for the best piece of white work in the whole Exhibition'.[78] The Depôt therefore provided disabled women with an arena in which they could not only participate in mainstream 'female' activities, but excel.

Sources Cited

Archival Sources

The Women's Library, London Metropolitan University, 5GFS/01/050.

The Women's Library, London Metropolitan University, 5GFS/02/002, 019, 334.

The Women's Library, London Metropolitan University, 5GFS/03/010, 014, 024, 025, 032, 052, 053, 054.

Printed Primary Sources

Census of England and Wales, 1901. (63 Vict. c. 4.). Summary tables. Area, houses and population; also population classified by ages, condition as to marriage, occupations, birthplaces, and infirmities, Parliamentary Papers, 1903 LXXXIV.1.

Every Woman's Encyclopaedia (London: Amalgamated Press, 1910–11).

Heath-Stubbs, Mary, *Friendship's Highway* (London: G.F.S. Central Office, 1926).

Reeves, Edith, *Care and Education of Crippled Children in the United States* (New York: Survey Associates, 1914).

Secondary Sources

Anderson, Anne, 'Victorian high society and social duty: the promotion of "recreative learning and voluntary teaching"', *History of Education*, 31/4 (2002): 311–34.

Borsay, Anne, *Disability and Social Policy in Britain Since 1750* (Basingstoke: Palgrave Macmillan, 2005).

Burnett, John (ed.), *Useful Toil: Autobiographies of Working People from the 1820s to the 1920s* (London: Routledge, 1994).

Callen, Anthea, *Women in the Arts and Crafts Movement 1870–1914* (London: Astragal Books, 1980).

Cheasley Paterson, Elaine, 'Crafting a national identity: the Dun Emer Guild, 1902–1908', in Betsey Taylor FitzSimon and James H. Murphy (eds), *The Irish Revival Reappraised* (Dublin: Four Courts Press, 2004), pp. 106–18.

Cumming, Elizabeth, 'Sources and Early Ideals', in Elizabeth Cumming and Wendy Kaplan, *The Arts and Crafts Movement* (London: Thames and Hudson, 1995), pp. 9–28.

Harrison, Brian, 'For Church, Queen and Family: The Girls' Friendly Society 1874–1920', *Past and Present*, 61 (1973), pp. 107–38.

Helland, Janice, *British and Irish Home Arts and Industries 1880–1914: Marketing Craft, Making Fashion* (Dublin and Portland, OR: Irish Academic Press, 2007).

Humphries, Steven, *Out of Sight* (Domino Films/Channel 4, 1992).

Humphries, Steve and Pamela Gordon, *Out of Sight: The Experience of Disability 1900–1950* (Plymouth: Northcote House, 1992).

LaCom, Cindy, '"It is More than Lame": Female Disability, Sexuality, and the Maternal in the Nineteenth-Century Novel', in David T. Mitchell and Sharon L. Snyder (eds), *The Body and Physical Difference* (Ann Arbor, MI: University of Michigan Press, 2004), pp. 189–201.

Mitchell, Patricia Mary Clyne, 'The Girls' Friendly Society 1875–1900', unpublished PhD thesis (Open University, 2003).

Mumm, Susan, 'Women and philanthropic cultures', in Sue Morgan and Jacqueline de Vries (eds), *Women, Gender and Religious Cultures in Britain, 1800–1940* (Abingdon: Routledge, 2010), pp. 54–71.

Richmond, Vivienne, 'Stitching the Self: Eliza Kenniff's Drawers and the Materialization of Identity in Late-Nineteenth-Century London', in Maureen Daly Goggin and Beth Fowkes Tobin (eds), *Women and Things, 1750–1950: Gendered Material Strategies* (Farnham and Burlington, VA: Ashgate, 2009), pp. 43–54.

Richmond, Vivienne, '"It is Not a Society for Human Beings but for Virgins": The Girls' Friendly Society Eligibility Dispute 1875–1936', *Journal of Historical Sociology*, 20/3 (2007), pp. 304–27.

Rose, Clare (ed.), *Clothing, Society and Culture in Nineteenth-Century England* (3 vols, London: Pickering and Chatto, 2011).

Schweik, Susan M., *The Ugly Laws: Disability in Public* (New York and London: New York University Press, 2009).

Stiker, Henri-Jacques, *A History of Disability* (Ann Arbor, MI: University of Michigan Press, 1999).

Notes

1 The Women's Library (TWL), 5GFS/02/019, Central Office – Press Cuttings Book.

2 Anne Borsay, *Disability and Social Policy in Britain Since 1750* (Basingstoke, 2005), p. 123.

3 Although some of the Society's patrons were men, it was otherwise an all-female organization. In using the potentially contentious term 'disabled' I am borrowing the Union of Physically Impaired Against Segregation's 1976 definition of disability as 'the disadvantage or restriction caused by a contemporary social organization which takes no or little account of people who have physical impairments and thus excludes them from the mainstream of social activities'. This I believe to be appropriate for the late nineteenth- and early twentieth-century context if slightly amended to: 'excludes them from mainstream activities'. Borsay, *Disability and Social Policy*, p. 10. I also recognize that the collective 'disabled women' imposes a fallacious homogeneity on a wide variety of experiences.

4 See, for example, the seminal Henri-Jacques Stiker, *A History of Disability* (Ann Arbor, MI, 1999).

5 The records form part of the GFS archive at The Women's Library, London Metropolitan University, and comprise correspondence between the Invalid Members, their Associates and the Depôt; application forms and a few needlework samples; promotional material; minute books and miscellaneous items. The Society's annual reports and various magazines also include information about the Depôt. In 1933 the Depôt was renamed Townsend Needlework (after the Society's founder, Mrs Townsend), but for clarity I refer to it throughout as the Depôt.

6 John Burnett (ed.), *Useful Toil: Autobiographies of Working People from the 1820s to the 1920s* (London, 1994), p. 129.

7 For histories of the GFS see Brian Harrison, 'For Church, Queen and Family: The Girls' Friendly Society 1874–1920', *Past and Present*, 61 (1973), pp. 107–38; Patricia Mary Clyne Mitchell, 'The Girls' Friendly Society 1875–1900', unpublished PhD thesis (Open University, 2003); Vivienne Richmond, '"It is Not a Society for Human Beings but for Virgins": The Girls' Friendly Society Eligibility Dispute 1875–1936', *Journal of Historical Sociology*, 20/3 (2007), pp. 304–27. The age of consent in England was raised from 12 to 13 in 1875, the year the GFS was founded. It was raised to 16 in 1885, the GFS playing a part in the amendment. Richmond, '"It is Not a Society"', p. 310.

8 There were also sister societies in Scotland, Ireland and overseas.

9 Mary Heath-Stubbs, *Friendship's Highway* (London, 1926), p. 63.

10 TWL, 5GFS/03/010, *Report of the Work and Progress of the Girls' Friendly Society*, 1885, p. 10.

11 TWL, 5GFS/03/014, *Report of the Work and Progress of the Girls' Friendly Society*, 1889, p. 11.

12 TWL, 5GFS/03/024, *Report of the Work and Progress of the Girls Friendly Society*, 1899, pp. 93–4.

13 TWL, 5GFS/02/002, *Wonderful Workers: The Story of the G.F.S. Central Needlework Depot*, pp. 3–5; TWL, 5GFS/02/002, royal correspondence; TWL, 5GFS/02/334, *Townsend House*, p. 6.

14 TWL, 5GFS/03/032, *Report of the Work and Progress of The Girls' Friendly Society*, 1907, p. 15.

15 TWL, 5GFS/03/052, *Report of the Work and Progress of The Girls' Friendly Society*, 1936, pp. 13, 29; TWL, 5GFS/03/053, *Annual Report and Statement of Accounts for the Year ended December 31st, 1940*, p. 16; TWL, 5GFS/03/053, *Annual Report and Statement of Accounts for the Year ended December 31st, 1941*, p. 15.

16 TWL, 5GFS/03/054, *Annual Report and Statement of Accounts for the Year ended December 31st, 1945*, p. 16.

17 TWL, 5GFS/01/050, Townsend Needlework Committee Minute Book.

18 *Every Woman's Encyclopaedia* (London, 1910–11), p. 2464; Anne Anderson, 'Victorian high society and social duty: the promotion of "recreative learning and voluntary teaching"', *History of Education*, 31/4 (2002), pp. 311–34. Anthea Callen, *Women in the Arts and Crafts Movement 1870–1914* (London, 1980), pp. 5–6; Janice Helland, *British and Irish Home Arts and Industries 1880–1914: Marketing Craft, Making Fashion* (Dublin and Portland, OR, 2007), pp. 8–10.

19 Helland, *British and Irish Home Arts and Industries*, p. 2; Elizabeth Cumming, 'Sources and Early Ideals', in Elizabeth Cumming and Wendy Kaplan, *The Arts and Crafts Movement* (London, 1995), pp. 9–28; pp. 19–21.

20 Helland, *British and Irish Home Arts and Industries*, p. 2. Cumming cites a precedent for the craft revival movement in the craft workshops established in mid-century to alleviate the poverty resulting from the Highland Clearances. The workers, she says, 'were, on the whole, female, and their disciplines were ones traditionally executed by women'. Cumming, 'Sources and Early Ideas', p. 19.

21 Helland, *British and Irish Home Arts and Industries*, p. 10. It is pertinent that in connection with the significance of women's magazines, Helland specifically cites *The Lady*. Ibid., p. 5.

22 TWL, 5GFS/03/025, Associates List for 1901.

23 Heath-Stubbs, *Friendship's Highway*, p. 64.

24 Helland, *British and Irish Home Arts and Industries*, pp. 12–13.

25 TWL, 5GFS/03/010, *Report*, 1885, p. 10; TWL, 5GFS/02/002, *Wonderful Workers*, p. 4.

26 TWL, 5GFS/03/025, *Report of the Work and Progress of The Girls' Friendly Society*, 1900, p. 96; Heath-Stubbs, *Friendship's Highway*, p. 64. According to Anne Anderson, some HAIA classes 'catered for the disabled', and left their legacy in present day … occupational therapy', but she gives no further information. Anderson, 'Victorian high society and social duty', pp. 317, 333.

27 Callen, *Women in the Arts and Crafts Movement*, p. 5.

28 Heath-Stubbs, *Friendship's Highway*, p. 63.

29 Helland, *British and Irish Home Arts and Industries*, p. 3.

30 Ibid., pp. 11, 28, 36–7.

31 Ibid., p. 3.

32 TWL, 5GFS/02/002, *Wonderful Workers*, p. 4.

33 Ibid., p. 7 and royal correspondence.

34 Callen, *Women in the Arts and Crafts* Movement, p. 5. For royal patronage of the HAIA see Anderson, 'Victorian high society and social duty', p. 325.

35 Harrison, 'For Church, Queen and Family', pp. 121–2; Anderson, 'Victorian high society and social duty', pp. 315–16.

36 TWL, 5GFS/02/002, application forms.

37 Susan M. Schweik, *The Ugly Laws: Disability in Public* (New York and London, 2009), p. 145.

38 Cindy LaCom, '"It is More than Lame": Female Disability, Sexuality, and the Maternal in the Nineteenth-Century Novel', in David T. Mitchell and Sharon L. Snyder (eds), *The Body and Physical Difference* (Ann Arbor, MI, 2004), pp. 189–201; p. 190.

39 Steve Humphries and Pamela Gordon, *Out of Sight: The Experience of Disability 1900–1950* (Plymouth, 1992), pp. 100–102. Borsay, *Disability and Social Policy*, pp. 101–4.

40 Borsay, *Disability and Social* Policy, p. 13.

41 TWL, 5GFS/02/002, application forms.

42 *Census of England and Wales, 1901. (63 Vict. c. 4.). Summary tables. Area, houses and population; also population classified by ages, condition as to marriage, occupations, birthplaces, and infirmities*, Parliamentary Papers, 1903 LXXXIV.1, pp. 288, 292. The Census did not delineate other forms of disability.

43 Schweik, *Ugly* Laws, p. 16; Humphries and Gordon, *Out of Sight*, p. 126.

44 Borsay, *Disability and Social Policy*, p. 106.

45 Edith Reeves, *Care and Education of Crippled Children in the United States* (New York, 1914), pp. 67–8.

46 Borsay, *Disability and Social Policy*, pp. 54–5.

47 Reeves, *Care and Education of Crippled Children*, pp. 69–70, 88–9.

48 Heath-Stubbs, *Friendship's Highway*, p. 63; TWL, 5GFS/02/002, *Wonderful Workers*, p. 6.

49 Vivienne Richmond, 'Stitching the Self: Eliza Kenniff's Drawers and the Materialization of Identity in Late-Nineteenth-Century London', in Maureen Daly Goggin and Beth Fowkes Tobin (eds), *Women and Things, 1750–1950: Gendered Material Strategies* (Farnham and Burlington, VA, 2009), pp. 43–54; p. 45.

50 TWL, 5GFS/02/019, Press Cuttings Book.

51 Heath-Stubbs, *Friendship's Highway*, p. 63.

52 Borsay, *Disability and Social Policy*, pp. 138–9.

53 Heath-Stubbs, *Friendship's Highway*, p. 63.

54 TWL, 5GFS/01/050, Minute Book.

55 *Census of England and Wales. 1901*, pp. 203, 205. The figure for domestic servants excludes those working in hostels, lodging and eating houses.

56 Humphries and Gordon, *Out of Sight*, p. 118.

57 *Census of England and Wales, 1901*, p. 199. These figures exclude other women working in garment manufacture such as tailors, shirtmakers and seamstresses, milliners.

58 Clare Rose (ed.), *Clothing, Society and Culture in Nineteenth-Century England* (3 vols, London, 2011), vol. 1, p. xi.

59 Callen, *Women in the Arts and Crafts Movement*, p. 3.

60 TWL, 5GFS/02/002, correspondence.

61 TWL, 5GFS/01/50, Minute Book; TWL, 5GFS/03/054, *Annual Report, 1945*, p. 11.

62 Cumming, 'Sources and Early Ideals', p. 21.

63 TWL, 5GFS/02/00, *Wonderful Workers*, pp. 5–6.

64 Elaine Cheasley Paterson, 'Crafting a national identity: the Dun Emer Guild, 1902–1908', in Betsey Taylor FitzSimon and James H. Murphy (eds), *The Irish Revival Reappraised* (Dublin, 2004), pp. 106–18; p. 111.

65 Schweik, *Ugly Laws*, pp. 47–8.

66 Susan Mumm, 'Women and philanthropic cultures', in Sue Morgan and Jacqueline de Vries (eds), *Women, Gender and Religious Cultures in Britain, 1800–1940* (Abingdon, 2010), pp. 54–71; p. 66.

67 TWL, 5GFS/02/002, correspondence.

68 The concern about infection was not peculiar to the Depôt but prevalent in the clothing trade more widely. Rose, *Clothing, Society and Culture*, vol. 2, p. 79.

69 TWL, 5GFS/02/002, correspondence.

70 Mumm, 'Women and philanthropic cultures', p. 66.

71 TWL, 5GFS/02/002, correspondence.

72 Steven Humphries, *Out of Sight* (Domino Films/Channel 4, 1992).

73 TWL, 5GFS/02/002, correspondence.

74 Borsay, *Disability and Social Policy*, p. 123.

75 Cumming, 'Sources and Early Ideals', p. 19. Anderson, 'Victorian high society and social duty', p. 325.

76 Cheasley Paterson, 'Crafting a national identity', p. 117.

77 TWL, 5GFS/03/025, *Report*, 1900, p. 96.

78 TWL, 5GFS/03/032 *Report*, 1907, p. 15.

9

Crafting Suburbia: The Community as Craft Object

Lily Crowther

Introduction

This chapter suggests that buildings, streets, neighbourhoods and perhaps even cities can be usefully conceived of as craft objects. It investigates the craft skills inherent in the building trades, and considers how the metaphor of 'community building' can be literally reified in a built environment.

These ideas are explored through a case study of the suburbs of London in the late Victorian and Edwardian periods (c.1880–1914), a time of rapid change both in the urban environment and in the crafts. In the late nineteenth and early twentieth centuries, London grew at an astounding rate. While the population of the inner city began to fall, the majority of new Londoners were accommodated in suburbia. The building of the new suburbs employed huge numbers of skilled craftsmen – belying the stereotypically uniform appearance of suburban housing stock, the building trades are among the most difficult to industrialize. Both the suburban building boom and the emergent studio craft movement reflected new kinds of relationships between labour, manufacturing and consumers in an age of unprecedented wealth. In shaping the physical nature of the city's new suburbs according to both conscious and unconscious ideals, craftsmen employed in the building trades played an active role in determining and exemplifying the values that characterized suburban life.

At the same time, transformative developments were taking place in craft education in Britain, with changes in legislation allowing local authorities to establish technical education colleges to teach applied arts and engineering trades. These institutions were aimed primarily at the working adult population, in particular young artisans. In response to growing demand from employers, schools of arts and crafts in outer London added subjects such as house-painting, ornamental plasterwork, stained glass, bricklaying and even plumbing to their curricula. As well as reflecting local training needs, this was often done with the explicit aim of improving the quality of the local built environment. The 23 schools administered and funded by the London County

Council were among the most active in the country in promoting craft education in the 1890s and 1900s.[1] Art schools might seem a surprising venue for such prosaic subjects as building, but this underlines the fact that the trades are essentially grounded in craft skills, and furthermore it indicates how important the quality and appearance of the built environment were to Londoners of the period.

The educational approach of the municipal schools was underpinned by the Technical Instruction Act of 1889, which gave local councils the power to levy a penny rate for 'a more perfect system of Science and Art Teaching, more especially for the artisan classes'. Under this Act, London County Council established a Technical Education Board in 1903 to support and guide its art schools and polytechnics.[2] Many of the students would have studied for the central examinations in Building Construction set by the government's Science and Art Department, which were available at Elementary, Advanced and Honours stages in a variety of areas of specialism, from bricklaying to joinery and plumbing.[3] However, although the focus of the examinations was firmly on the technical aspects of construction, the artistic and design elements nevertheless also formed a major part of the curriculum on many courses. Surviving archival material from a small number of London colleges allows us to examine the relationship between the physical process of building the suburbs and the activities and aspirations of the schools' governing bodies and students.

This chapter uses this primary material to explore three main areas: what was being taught in the suburban art schools with regard to the building trades; the character of the student body; and the links between the students' work and the nature of the local built environment, and specifically its qualities as craft. Considering a building as a collective craftwork is not unprecedented, and there have also been several studies of the more decorative aspects of the building trades in the renaissance and early modern periods, dealing for example with ornamental plasterwork or stone-carving. However, the predominant contemporary conception of craft is as an individual creative pursuit. As Ezra Shales has noted in his study of the crafting of the Empire State Building: 'the meaning of craft narrowed in the late twentieth century to connote a heroic anti-establishment retreat from both conformity and mechanisation, and for many this definition is still in service'.[4] It could be argued that this narrowing of meaning took place much earlier, as the studio crafts were born in the late nineteenth and early twentieth centuries in response to the march of industrialization. At this period, for the first time, craft and the processes of hand-making became an active choice for practitioners of many trades, rather than the only way an object could be produced. But in counterpoint to this dominant idea of independent creative practice, notions of collective craft give us a way to think about community building and collaboration in an active sense, as also considered in Chapter 10.

Moving beyond the archival evidence, this chapter draws on the interdisciplinary methods of cultural geography and psychogeography to look at the roles of the emotions and the imagination in the construction of cities, through Steve Pile's exploration of the phantasmagoric aspects of city life, and Iain Sinclair's focus on London and its layers of meaning. Pile's and Sinclair's conceptions of the city as created, in many different senses, by its inhabitants are here used to re-imagine suburbia as a communal craftwork.

The Expansion of London's Suburbs

Having entered the nineteenth century with a population of over a million, London by 1901 was a megalopolis of over 6.5 million inhabitants.[5] The central reason for this expansion was the changing distribution of wealth in Britain – the Industrial Revolution and the new sources of revenue in the empire transformed patterns of working and investment in ways that radically affected urban life. The last quarter of the nineteenth century saw a steady rise in real incomes and a growth in the number of administrative, clerical and skilled manual jobs. As well as providing a salary that would allow workers to contemplate home ownership, or at least renting a whole house, these jobs also offered regular, relatively short hours. This meant that a long commute became a realistic option, since workers had enough leisure time to travel.[6] New suburban development provided an excellent investment vehicle in the form of ground rents and mortgages, at a time when relatively few alternatives were available.[7] British agricultural land was losing value as a consequence of imperial expansion, and residential construction also offered a way to give new value to that land. There was an inversely proportionate cycle linking residential building at home to investment abroad – when foreign investment opportunities were too limited or risky, British investors channelled their wealth into housing (hence the phrase 'safe as houses').[8]

In the late nineteenth and especially the early twentieth centuries, London's public transport systems were also transformed. Improvements in the speed and accessibility, both physical and financial, of road and especially rail transport gave the residents of ever greater areas the opportunity to travel regularly across London (see Figure 9.1). The expansion of railway networks in particular was often linked directly to the construction of 'railway suburbs', planned neighbourhoods built in areas newly linked or about to be reached by trains; the developments were often motivated by land speculation by the railway companies.[9]

All of these factors led to striking growth not just in the city as a whole, but specifically in the suburbs. Contemporary journalist and historian Sidney Low, analysing the census returns for the period 1881–91, noted that 'contrary to popular belief, Britain's population was growing "not in the cities themselves, but in the ring of suburbs which spread into the country"'.[10] Britain's top four areas of population growth in the 1880s were all lower-middle-class London railway suburbs, and from 1861 until 1901, the population of the outer ring of London's suburbs grew by about 50 per cent each decade, reaching almost 2 million; meanwhile, the population of inner London steadily declined.[11]

Much of the new building in the suburbs was carried out by speculative builders. This new breed of developers would lay out a street, a block or just a few houses at a time. The most influential and prolific speculative builder was Thomas Cubitt (1788–1855); while earlier builders had subcontracted much of their work to independent specialists such as masons, carpenters and plasterers, Cubitt employed his own craftsmen and labourers in a single firm, thus setting the model for later speculative builders across the city.[12] But most builders operated on a much smaller scale than Cubitt, and many were completely untrained, entering the building trade purely in search of an easy profit – they included

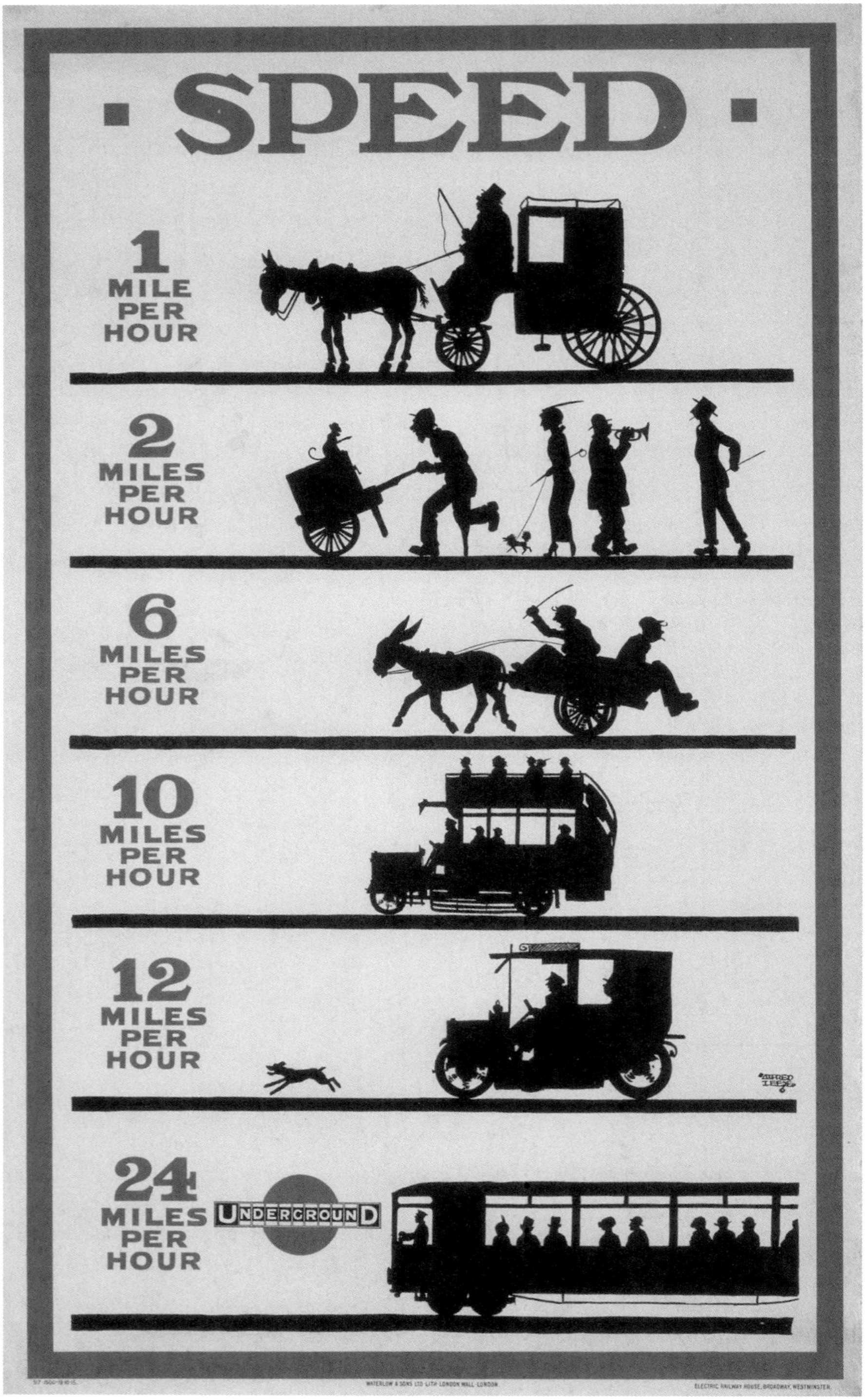

9.1 *Speed*, poster for London Underground, Alfred Leete, 1915.

'labourers and mechanics, servants and publicans, shopkeepers and merchants, lawyers and clergymen'.[13]

Trade Skills as Crafts: Teaching the Building Trades in London's Suburban Art Schools

Naturally this explosion in suburban building created a huge demand for skilled tradesmen. This in turn stimulated a burst of training activity across the capital, and courses were set up for young people, apprentices, artisans and professionals wishing to develop or expand their knowledge of this potentially lucrative field. One reason why training was particularly crucial was that, unlike other forms of production, the building trades are remarkably difficult to mechanize. At a time when other traditional crafts were being undermined by the efficiency of factory production, skills such as masonry, bricklaying, plumbing, plastering and the many other specialist trades required to build a home remained essential – their 'site-specific', case-by-case nature meant that they could not be replaced by unskilled labour.

Suburban housing was often condemned by Victorian and Edwardian commentators for its monotony. As Olsen observes:

> The inherent internal contradiction in a society that glorified individuality yet required standardisation, that longed to re-establish lost patterns of craftsmanship yet depended on the machine, that tried to reconcile a concern for the well-being of the mass of the people with aesthetic aims that could only be realised at an expense that would limit their products to a wealthy few, is not exhausted by an examination of its attitude towards suburban housing.[14]

Yet the suburban house is a revealing case study, combining a reputation for the most stultifying, characterless repetition with a genuine local variability and dependence on the contributions of highly skilled individuals. Howard Davis has explained this in terms of the skilled interpretation of implicit rules: rather than working with detailed design drawings and prescriptive building regulations, as is normal today, the Victorian builder exercised his judgement in applying his knowledge of acceptable building types to the specific circumstances at hand.[15] For example, a speculative builder might refer to a manual or pattern book of standard schematic house designs, or have in mind an ideal design of his own:

> but faced with the context of an oddly sloping piece of land, or an irregularly shaped site, or the need to preserve light to a neighbour's window, or available roof timbers of a certain length, [he] must call on precise discretionary judgment in order to transform those chosen types into a real building.[16]

This required possession of a huge body of largely uncodified, flexibly applicable, tacit knowledge, which could not be transferred to an off-site industrial context or split into less skilled production-line tasks – Davis describes this as a system of 'rules [which] were largely inherent in the craft itself'.[17] In turn, this embodied knowledge and judgement resulted in the typical variety of the urban landscape, which emerged from the builders' idiosyncratic application of their craft skills.

Formal training in the building trades was offered through the emerging system of local art schools supported under the Technical Instruction Act. Such schools were established across the country, each tending to focus its curriculum on subjects of relevance to local industries, such as pottery in Stoke-on-Trent, or textile design in Manchester. Those in London enabled increasing numbers of students to develop the required skills to participate in the city's building boom. Among the new schools were Hornsey School of Art, opened in 1880 (now part of the University of Middlesex), Battersea Polytechnic Institute (see Figure 9.2), opened in 1894 (now the University of Surrey), and Camberwell School of Arts and Crafts, opened in 1898 (now Camberwell College of Arts, part of the University of the Arts London). These three are unusual among London schools of the period in that they have rich surviving archives. They also span a large area of the city, representative of the developments of the time: Hornsey School was sited in the northern district of Crouch End, which remained rural until around 1880, but which was served by seven railway stations by 1887, making it a commuter hub; Battersea is one of the inner western suburbs, just south of the river Thames, and was already a much more industrial area with several railway lines running through it from the 1830s onwards; and the neighbourhood of Camberwell is in the south-east, and had been a popular rural resort for city-dwellers until the arrival of the railways in the 1860s precipitated a population explosion. An exploration of the teaching of the building trades at these three schools therefore offers a good cross-section of wider practice in the capital in this period.

Hornsey was a relatively small school, with a consequently narrow curriculum, strongly focused on drawing and modelling. As the headmaster explained in 1911:

> The Evening classes are attended by Designers, Draughtsmen, Artists, Book decorators, Teachers, Lithographers, Carvers, Metal Workers, Sign writers, and by students occupied in various Art Industries, such as Textile and Wall paper Manufacture, Stained Glass, Furniture and House Decorations, but as the number who attend in any one of these various Art Industries are small, it is not possible to plan separate courses of work for them.[18]

Yet, although the scope for specialist teaching at Hornsey was limited, clearly efforts were made by the staff to support artistic education for the building trades. The local newspaper noted the introduction of 'machine and building construction and architectural drawing' to the curriculum in 1884.[19] In 1907, the school requested additional funding from the county council to expand its premises, including the construction of a classroom for 'Painters' and Decorators' Work'.[20]

The situation at Battersea was somewhat different, because as a polytechnic it was much larger and had several departments, including those covering scientific and technical subjects as well as art. Most subjects relating to the building trades were taught in a combination of technical and artistic contexts, with the balance of each specific course varying according to the particular requirements of the trade. The focus in the Art Department as a whole was firmly on the applied arts, particularly in relation to the building trades. According to the first prospectus: 'The general scheme of Art Instruction will comprise a thoroughly practical knowledge of Designing, Drawing, Painting, and Modelling, especially in its various applications to TRADES and INDUSTRIES (THE BUILDING TRADES, DECORATING and PAINTING, MOSAIC… &c,

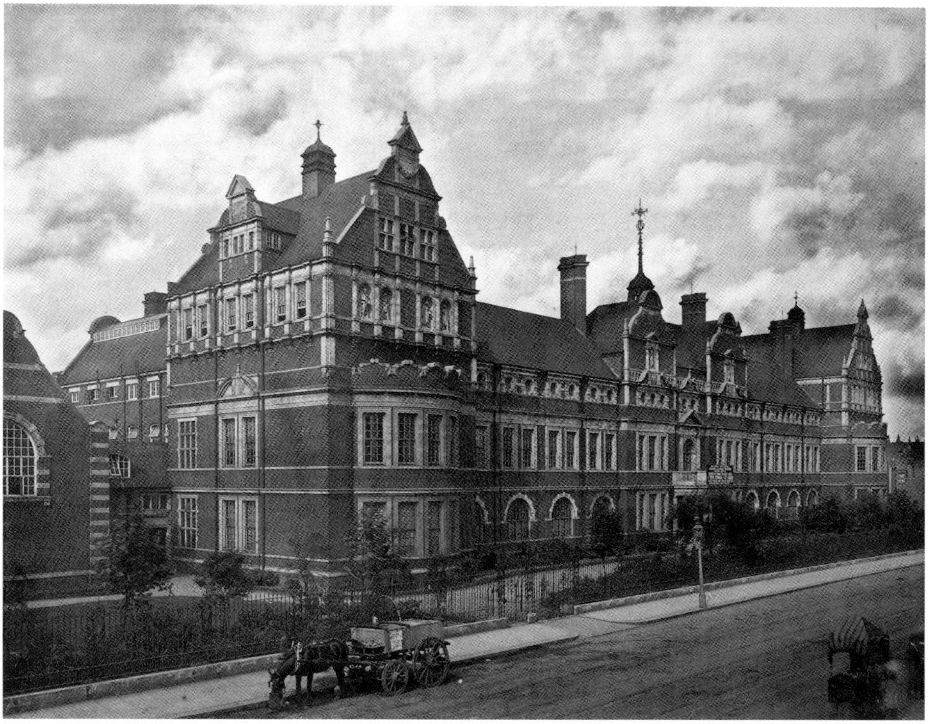

9.2 Battersea Polytechnic Institute, architect E.W. Mountford, 1894.

&c)'.[21] From its foundation, the school offered a wide range of introductory courses in 'Mechanical Engineering and Building Trades', ranging from 'Carpentry and Joinery' to 'Strength and Properties of Materials of Construction', as well as more advanced specialist trade classes such as 'Plasterer's Work' and 'Electric Wiring and House Fitting'.[22] There was also a lengthier programme of study available in 'Building Construction and Drawing', including 'Principles of Planning and Designing Dwelling Houses'.[23] By 1895, the prospectus recommended extended programmes of study for those engaged in a wide variety of trades, combining classes in the areas of 'Science' (such as 'Practical Geometry', 'Builders' Quantities' and 'Theoretical Mechanics (Fluids)'), 'Trade' (for example 'Electrical Workshop'), and 'General' (including 'Magnetism and Electricity', 'Sound and Light' and 'Hygiene'), as well as art classes in 'Freehand Drawing and Design'.[24]

Gradually more classes were introduced, including subjects relating to the design of broader urban environments, rather than simply individual buildings. For example, in 1896 a course was established in 'Land Surveying and Road Making', including 'Surveys for New Works, covering New Roads, Railway Lines, Water Supply, Drainage and Sewerage Works, Tunnels, Contouring'.[25] A new course in 'Illuminating Engineering' began in 1911, with the prospectus explaining that:

> The consideration of the most suitable methods of lighting Public Buildings, Schools, Streets, &c, is of the greatest importance, and too much stress cannot be laid upon the necessity for a deeper study of the subject by those engaged in such work.[26]

It is particularly interesting to think in this context about the design of whole streets and neighbourhoods in terms of craft skills – building a new suburb involved the sensitive and thoughtful provision of public amenities and utilities, not just an expanse of housing.

Another key development at Battersea was the increasing sophistication of the architecture curriculum. This began as a relatively small element of 'Building Construction and Drawing', with the syllabus for the most advanced students culminating in 'Design – a small detached residence'.[27] From 1898, a course of lectures in English architecture was offered, with reduced fees for any student enrolled in a building or art class.[28] By 1903, the school provided a two-year programme of day classes in 'Architectural and Constructional Work' for young students, including engineering, physics, mechanics and art.[29] For the older artisans, there was an architecture course including lectures in architectural history and ornament, study visits and drawing from historical buildings, looking particularly at 'the influence of constructive conditions and materials upon Architectural styles' and aimed chiefly at students of building construction.[30] Eventually, architecture became a full three-year programme integrating science, construction, surveying, architectural history and design.[31] This development represents a broadening of access to the architectural profession, with an emphasis on the development of craft skills rather than simply abstract design skills.[32]

At Camberwell School of Arts and Crafts, the building trades were also central to the curriculum from the start. The prospectus in the opening year explained that: 'The instruction is adapted to the needs of those engaged in the different departments of Building Work (Architects, Builders, Modellers, Decorators)'.[33] Courses were offered in house-painting and decorating, plastering, masonry and stone-carving, and architecture.[34] The 1900 prospectus elaborates on the content of the 'Architectural Design' course: 'The subject is treated ... from the point of view that architecture should respond to the facts of modern life. Demonstration of the mechanics of construction will be given concurrently with instruction in design'.[35] The class was successful, but many potential students were put off because it required a prior knowledge of building, so a 'Building Construction' course was offered to supply this need.[36] In 1903, the school was expanded with new facilities for more practical, workshop-based teaching in a range of subjects, including brickwork, carpentry, plumbing, wrought-ironwork, and stained glass.[37]

Who were the Students?

Whilst considerable detail on curricula can be found in prospectuses and elsewhere, it is much more difficult to ascertain who was taking these courses. However, some reasonable guesses at the general character of the student body can be made by examining the stated aims of each institution, the restrictions placed on entry to some courses, and the incentives offered to potential students. On opening in 1898, Camberwell described its aims as: 'To give the best artistic and technical education to all classes in the district. To supplement the knowledge gained by craftsmen in workshops. To help the craftsman to become the designer of his work'.[38] Although claiming to support 'all classes', this clearly suggests that the primary target audience was composed of artisans. Many of

Camberwell's classes were specifically aimed at apprentices or workers in a particular trade; for example the course in house-painting was 'intended for apprentices and artisans engaged in the Painting and House Decorators' Trade'.[39]

Battersea Polytechnic aimed many of its courses at a similar audience, although usually dividing them more finely according to level of experience. For example, its day classes in 'Mechanical Engineering and Building Trades' were 'intended for Apprentices or other young Artisans who may be able to obtain necessary absence from their work'.[40] Some courses offered scholarships for students under the age of 18, such as 'Painters' and House Decorators' Work',[41] while specialist practical classes in particular trades such as 'Brickwork and Masonry' or 'Plumbers' Work' were 'only open to persons actually working in the trades with which they are connected'.[42] The Polytechnic later introduced more academic courses for students intending to sit professional examinations, or for those with a good school education who wished to prepare for workshop or office training in 'Architectural and Constructional Work'.[43] These sat alongside the 'Technical Day School' for young students with elementary school education preparing for an apprenticeship.[44]

A more detailed snapshot of a single cohort of students is given by a loose-leaf manuscript sheet inserted into the front of the minute book of Hornsey College of Art for the period 1901–15; the sheet itself is undated but presumably comes from this period. It is headed 'Occupations of Students (or Parents)', and lists 46 categories totalling 264 individuals. Of these, two are listed as 'Stained Glass', two 'Stone Carvers', three 'Architects', three 'Engineers', one 'Decorator' and one 'Builder'; 118 are 'Students', presumably meaning young people in full-time education, so that, of the working adults attending the college, just over 8 per cent were engaged in the building trades.[45] The school's annual report for the session 1905–06 noted: 'It is gratifying to record that the School attracted a large number of students engaged in various Artistic trades such as Architects, Designers, Draughtsmen, Illustrators, Embroiderers, Decorators, Glass and Metal Workers etc'.[46] Since Hornsey offered predominantly general courses in drawing, painting and modelling, it is probable that the proportion of students involved in the building trades was higher at other institutions such as Battersea and Camberwell, which offered specialist tuition in these trades. Overall, this brief survey suggests that the majority of students in subjects relating to the building trades were drawn from the upper working and lower middle classes – those equipped with either specialist craft skills, or the academic ability and ambition to enter the world of building design, if not of professional architecture.

The students were drawn not just from the immediate locality of each school, but from surrounding districts, illustrating the interconnectedness of the suburbs. For example, the students of Hornsey School of Art in 1899 travelled from across a substantial area of north London (see Figure 9.3).[47] Battersea Polytechnic advertised its convenient transport links at the front of every prospectus. These connections between suburbs are often overlooked in the concentration on links between suburb and centre. As Alan Jackson points out:

> The important transport routes were not just the radial services, but those linking the suburbs to one another, and thus giving them a life of their own; people could commute between suburbs, and travel to shops, without being dependent on the city.[48]

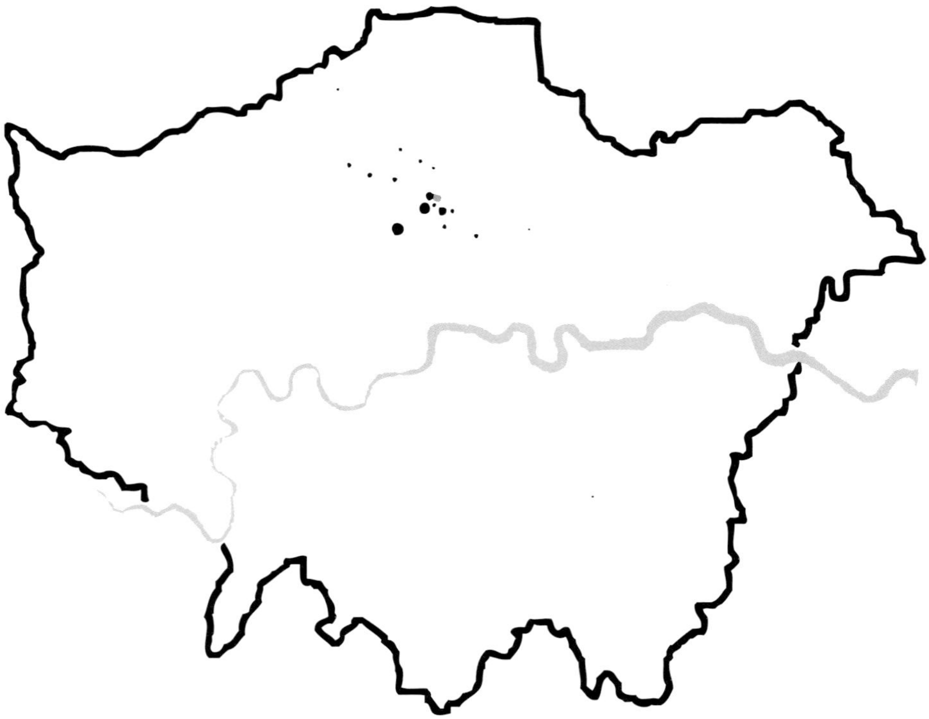

9.3 Map of Greater London showing home districts of students at Hornsey School of Art in 1899. Data drawn from Hornsey School of Art, *Seventeenth Annual* Report, 1899.

The size of each dot is proportional to the number of students from that area. The school is indicated by the lighter dot.

Many suburban residents worked locally, in both service and manufacturing industries.[49] The suburbs were not mere dormitories, but hubs in their own right – centres of economic activity, education and community.

Handmade Neighbourhoods: The Built Environment as Craft Object

Considering the skills of those engaged in the building trades in terms of craft leads naturally to the idea that the products of their skills – the built environment – constitute craft objects. They are large, complex and difficult to pin down, very different from the 'medium-sized dry goods' (in J.L. Austin's phrase) that constitute what we usually describe as craft objects, but thinking about buildings in this way is nevertheless useful. It offers a new perspective on the relationships between people and environments, moving beyond the relatively abstract architect–client dialogue to encompass a more active, collaborative idea of community-making.

It is clearly difficult in many ways to extend the idea of the built environment as craft object beyond a single building, or planned group of buildings, to a wider neighbourhood.

Not only is work on a neighbourhood distributed between many craftspeople (as it usually is on a single building), but there are also multiple designers, often working without direct reference to one another. However, to consider the whole neighbourhood as in some senses a single collaborative craftwork is a useful strategy, as it allows us to frame questions about community values, which may be unconscious but are still shared between fellow residents. This is where the roles of the land surveyor, the street lighting expert and the landscape gardener come especially strongly into play. Their work can be seen as a realization of the abstract collective ambitions held by residents of a wider neighbourhood, reaching beyond the desire simply for a commodious private home and access to the city, and taking in ideas of community and mutual obligation.

Mary Corbin Sies has put forward a potential ethnographic approach to the question, 'What meanings did suburbia hold for those who were its creators and original consumers?' She proposes looking at architecture in a performative sense, to 'reveal the dynamic nature of the relationship between suburbanites and their material culture'.[50] Corbin Sies's case study looks at a small number of architect-designed residences in upper-middle-class American suburbs, but she notes that:

> professional men did not dominate the suburban discourse. It is more accurate to view architects – many of whom … made their homes in these suburbs or ones like them – as fellow members of the social stratum that composed each suburb's community; … the communities that created the suburban ideal conceptualized a thorough environmental program that encompassed not only the home but also … the larger suburban community. Any examination of the suburban ideal must acknowledge the interconnectedness of the suburban landscape.[51]

Although her work focuses on the upper-middle-class American context, Corbin Sies's ideas are readily applicable to other types of suburban communities; her ethnographic approach works well when considering individual buildings designed by known architects for named clients, but can be adapted to take a more design-historical or social-historical perspective when considering larger areas in the context of largely anonymous speculative building.

The idea that suburbs embody the values of their communities has been widely discussed, particularly in the American context. Dolores Hayden discusses the suburb as a middle-class phenomenon – while the rich had both country and city houses, and the poor tended to move directly from farms to the inner city, the middle classes:

> tried to have it all, to sustain a country ambience near the city with just one residence. They struggled with the difficult commutes and became the most ardent defenders of building single-family houses among the trees and flowers, removed from the pollution, epidemics and economic stresses of the city … they invented a new approach to residential space located on the borders.[52]

Hayden looks at a range of suburban development types, of which the most relevant here is the 'picturesque enclave'. Built around American cities from the 1850s onwards, these were characterized by 'curving roads', 'heavy planting', 'shared parks and other common spaces'; according to Hayden:

> the designers of enclaves added a sense of community to the borderland goals of house and land, becoming the first to express a triple dream. Picturesque enclaves were the most important secular

> manifestation of a wider communitarian movement whose adherents believed that building a model community in a natural setting led to the reform of society.[53]

In the American suburb, self-building was much more common than in Britain, creating a more diverse streetscape. Buyers usually had to conform to various regulations regarding set-back distances, boundary fencing, and sometimes even building materials and styles, but were otherwise free to be much more expressive than was possible in the regimented environment of the inner city. Quoting a consumer magazine from 1917, Robert M. Fogelson explains:

> Suburbia encourages 'individualism', one suburbanite wrote. There 'you may wear what you please', she said, and do other things 'you would never dare in the City'. There 'you may, if you wish, paint your house orange and purple and put a pink roof on'.[54]

In London's Victorian suburbs the practice of speculative building was much less conducive to this kind of individualism. Not only were houses constructed in larger groups, but they needed to appeal broadly to potential buyers or tenants. They might therefore be thought of as more representative of shared ideals than the individualist self-build model. Their desirability lay in their capacity to fulfil people's wishes for community as well as for house and land – as described by Hayden, the suburb offered the possibility of a new kind of life. It embodied the values of the growing middle class, values associated with privacy and family, as well as respectability and contributing to the welfare of the community, in an architecture that borrowed heavily from vernacular traditions to evoke the image of an old-fashioned way of life (see Figure 9.4). This suburban landscape of separate houses, each with its own garden but sharing communal spaces such as parks, evidenced a different set of values to those demonstrated by other late nineteenth-century streetscapes, such as those of Haussmann's Paris.[55]

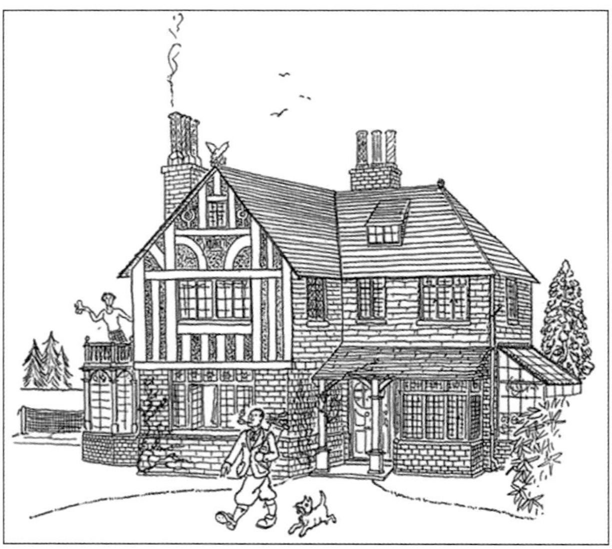

9.4 'Wimbledon Transitional', Osbert Lancaster, from *Pillar to Post*, 1938.

Looking specifically at the physical making of the suburbs, or any other built environment, offers a way of explaining how this metaphorical embodying of values takes place in a real context. The idea of literally constructing a community can be more or less consciously deployed by developers, as Shales has observed in the context of the Empire State Building:

> [the building's] stability is rooted in ... the idea that ... one and all, the speculator and architect, [the politician] Al Smith and the day labourer, can ally as civic-minded craftsmen in the service of a larger and a greater collective good.[56]

However, it is more often unconscious, underlying the collective decision-making of builders and buyers alike, and conditioning the ways that neighbourhoods develop.

In London, this process was in part underpinned by the efforts of the local authorities to provide a suitable education in the building trades. In some cases, the art schools had a very direct influence on the character of their local urban fabric. For example, in 1900, 28 new municipalities were created in London, taking over the duties of local vestries under the London Government Act. New council chambers were built across the city to house the new municipal government bodies. In Camberwell, the *News of the World* described how:

> the students of the School of Arts and Crafts are preparing frescoes illustrative of local history, which are to adorn the panels of what next month will be the council-chamber. The students in question are those who have obtained scholarships from the London County Council, and they are being encouraged to leave permanent evidence of their skill on the public buildings of the district.[57]

The largest of the trade classes at Camberwell School was the house-painting and decorating course, in which the staff noted:

> Efforts are being made to lead the students away from the worst features of the Decorators' Trade and aim at a higher class of work. Trade prejudice is strong but some success has been achieved.[58]

This remark suggests an explicit aim among the staff to use the school as a means to improve the local environment. In a more general sense, it must be assumed that the atmosphere of the art schools, with their strong links to the Arts and Crafts movement, contributed to the prevailing pseudo-vernacular styles of London suburbia.

Conclusion: Realizing the Collective Imaginary

Thinking about the literal crafting of the suburban environment offers a new perspective on the metaphorical idea of community-building, and raises questions about how a city is made. Once we understand building as a kind of collective craftwork, we can see that concepts usually associated with the creative mind, such as imagination, emotion and intuition, are real and substantial elements in the material spaces of the city. Steve Pile premises his *Real Cities* on the idea that 'to deal with the real city – the city as it really is – does not just mean taking into account its physical and social structures, but also its mental life'.[59] For Pile, 'What makes the city a city is not only the skyscrapers or the shops or the communication networks, but also that people in such places are forced to *behave* in *urban* ways'.[60] We might propose that many of the characteristically 'urban ways' of behaving relate to the construction and ongoing negotiation of the built environment – when almost everything about one's surroundings is manmade, and densely occupied, the precise character of buildings and their interrelationships becomes the source of much tension and psychological focus.

Pile describes cities as 'part of the machinery of desire: they are sites where dreams are forged, housed, circulated'; 'cities … seem to be the product of human wish-fulfilment'.[61] Utopian writings such as Ebenezer Howard's *Garden Cities of Tomorrow* (1902) use the city as the arena for the construction of a perfect society, 'perhaps because cities appear to be the product of pure human endeavour'.[62] Pile frames his ideas in terms of the emotional work involved in producing and reproducing city spaces, and the resultant emotional qualities of those spaces, but it seems useful to generalize his ideas and apply them in more literal, practical ways. His focus on the lived experience of space offers a new perspective on what he calls the 'geographical under-labourings that create the seemingly unremarkable spatialities of city life', an apt description of the unsung contributions of the anonymous craftsman.[63]

Pile is interested in the psychogeographical writings of Iain Sinclair, which he describes as aiming 'to provide an accurate assessment of the mood of London'.[64] He sees in Sinclair's investigation of 'the amnesias, frustrations and diversions of the city' an exposition of the way in which 'the times and the spaces of the city are … "over-determined" (ie determined many times over): by memory, by meaning, by social relations'.[65] Ged Pope notes:

> an intriguing orbicular expansion in the focus of Sinclair's work, analogous to the restless outward growth of the suburbs, from his earlier work's concern with mapping the urban centre, toward an increasing need to survey the suburban margins.[66]

In his *London Orbital*, Sinclair takes this need to its extremes, walking the route of the M25 orbital motorway around London's outer suburbs. He describes his journey 'through the suburbs at night … trying to find some way to unravel the syntax of London'.[67] The syntax of a city – the way its elements come together to create new meanings – is precisely what is being constructed when a new neighbourhood is planned and built. In the Victorian and Edwardian suburbs, built at such speed and with consequently strong coherence of style, the syntax of houses, streets and public spaces allows us an insight into the values of the communities who made them. The mood of this particular side of London came from a complex web of new social relations and economic realities, which its inhabitants attempted to ground in shared (illusory) memories of an idyllic past, when craft traditions and community values were more meaningful. Through their desires and skills, they were able to realize a collective wish-fulfilment of sorts, which still shapes vast tracts of the city today.

Acknowledgements

This chapter builds on several years of work which have been supported by many colleagues at the Victoria & Albert Museum, particularly Glenn Adamson and Angela McShane. I would also like to thank the archivists who gave me access to their collections: Gustavo Grandal-Montero at Camberwell College of Arts, Kay Butcher at the University of Surrey, and Judy Vaknin at Middlesex University. Finally, I am grateful to staff at the Material Culture Institute, University of Alberta, for the opportunity to present my research at their 2011 conference, and to all those who offered suggestions on that occasion, as well as to the editors of this volume for their comments.

Sources Cited

Adams, Henry, *Building Construction: Key to Examinations of Science and Art Department* (London: Chapman & Hall, 1894).

Ball, Michael and David Sutherland, *An Economic History of London 1800-1914* (London: Routledge, 2002).

Barson, Susie, 'Infinite Variety in Brick and Stucco: 1840–1914', in Andrew Saint (ed.), *London Suburbs* (London: Merrell Holberton, 1999).

Battersea Polytechnic Institute, *Prospectus of Day and Evening Classes, Session 1894–5* (London: Battersea Polytechnic Institute, 1894).

Battersea Polytechnic Institute, *Prospectus of Day and Evening Classes, Session 1895–6* (London: Battersea Polytechnic Institute, 1895).

Battersea Polytechnic Institute, *Prospectus of Day and Evening Classes, Session 1896–7* (London: Battersea Polytechnic Institute, 1896).

Battersea Polytechnic, *Prospectus of Day and Evening Classes, Session 1898–9* (London: Battersea Polytechnic, 1898).

Battersea Polytechnic, *Prospectus of Day and Evening Classes, Session 1903–4* (London: Battersea Polytechnic, 1903).

Battersea Polytechnic, *Prospectus of Day and Evening Classes, Session 1904–5* (London: Battersea Polytechnic, 1904).

Battersea Polytechnic, *Prospectus of Day and Evening Classes, Session 1909–10* (London: Battersea Polytechnic, 1909).

Battersea Polytechnic, *Prospectus of Day and Evening Classes, Session 1911–12* (London: Battersea Polytechnic, 1911).

Camberwell School of Arts and Crafts, *Prospectus and Timetable* (London: London County Council Technical Education Board, 1898).

Camberwell School of Arts and Crafts, *Prospectus and Timetable* (London: London County Council Technical Education Board, 1900).

Camberwell School of Arts and Crafts, *Prospectus and Timetable* (London: London County Council Technical Education Board, 1903).

Camberwell School of Arts and Crafts, Joint Advisory Sub-Committee minutes, 7 May 1900 (unpublished, Camberwell College of Arts Library).

Corbin Sies, Mary, 'Towards a Performance Theory of the Suburban Ideal, 1877–1917', *Perspectives in Vernacular Architecture*, 4 (1991), pp. 197–207.

Crowther, Lily, 'Et in Suburbia Ego: A Cultural Geography of Craft in the London Suburbs', *Journal of Modern Craft*, 3/3 (July 2010), pp. 143–60.

Davis, Howard, 'Explicit Rules, Implicit Rules and Formal Variation in Vernacular Building', *Perspectives in Vernacular Architecture*, 4 (1991), pp. 53–63.

Dyos, Harold J., *Victorian Suburb: A Study of the Growth of Camberwell* (Leicester: Leicester University Press, 1961).

Fogelson, Robert M., *Bourgeois Nightmares: Suburbia, 1870–1930* (New Haven, CT: Yale University Press, 2005).

Hayden, Dolores, *Building Suburbia: Green Fields and Urban Growth, 1820–2000* (New York: Random House, 2003).

Hornsey School of Art, Minute Book of the Committee Meetings of the Hornsey School of Art, 1882–1901 (unpublished, Middlesex University Archive).

Hornsey School of Art, Minute Book of the Committee Meetings of the Hornsey School of Art, 1901–15 (unpublished, Middlesex University Archive).

Hornsey School of Art, *Prospectus* (London: Hornsey School of Art, 1905).

Hornsey School of Art, *Twenty-Fourth Annual Report* (London: Hornsey School of Art, 1906).

Jackson, Alan A., *Semi-Detached London* (Didcot: Wild Swan Publications, 1991).

Kuchta, Todd, *Semi-Detached Empire: Suburbia and the Colonization of Britain, 1880 to the Present* (Charlottesville, VA: University of Virginia Press, 2010).

Lee, R.G. (Honorary Secretary, Hornsey School of Art), letter to the Education Committee, Middlesex County Council, 11 February 1907 (unpublished, Middlesex University Archive).

News of the World, 'London Municipalities', 14 October 1900, 3.

North Middlesex Chronicle, 'Hornsey Government School of Art', clipping pasted into the Minute Book of the Committee Meetings of the Hornsey School of Art, 1882–1901, between entries for 30 October 1884 and 8 December 1884.

Olsen, Donald J., *The Growth of Victorian London* (London: Batsford, 1976).

Pile, Steve, *Real Cities: Modernity, Space and the Phantasmagorias of City Life* (London: Routledge, 2005).

Pope, Ged, 'Deep in South London', *Literary London: Interdisciplinary studies in the representation of London*, March 2008. Available at: www.literarylondon.org/london-journal.

Roach, John, *Secondary Education in England* (London: Routledge, 1991).

Shales, Ezra, 'Constructing the Empire State Building', *Journal of Modern Craft*, 4/2 (July 2011), pp. 119–45.

Sinclair, Iain, *London Orbital* (London: Penguin, 2003).

Swinstead, Frank H., 'Hornsey School of Art: Head Master's Statement of Organisation and Curriculum', 1911, in Hornsey School of Art, Minute Book of the Committee Meetings of the Hornsey School of Art, 1901–15 (unpublished, Middlesex University archive).

Ward, David, 'A Comparative Historical Geography of Streetcar Suburbs in Boston, Massachusetts and Leeds, England: 1850–1920', *Annals of the Association of American Geographers*, 54/4 (1964), pp. 477–89.

Notes

1 Lily Crowther, 'Et in Suburbia Ego: A Cultural Geography of Craft in the London Suburbs', *Journal of Modern Craft*, 3/3 (July 2010), pp. 150–51.

2 John Roach, *Secondary Education in England* (London, 1991), pp. 107–9.

3 For sample examination questions, see Henry Adams, *Building Construction: Key to Examinations of Science and Art Department* (London, 1894).

4 Ezra Shales, 'Constructing the Empire State Building', *Journal of Modern Craft*, 4/2 (July 2011), p. 140.

5 Michael Ball and David Sutherland, *An Economic History of London 1800-1914* (London, 2002), p. 42; Harold J. Dyos, *Victorian Suburb: A Study of the Growth of Camberwell* (Leicester, 1961), pp. 54, 19.

6 Alan A. Jackson, *Semi-Detached London* (Didcot, 1991), p. 9.

7 Donald J. Olsen, *The Growth of Victorian London* (London, 1976), p. 156.

8 David Ward, 'A Comparative Historical Geography of Streetcar Suburbs in Boston, Massachusetts and Leeds, England: 1850–1920', *Annals of the Association of American Geographers*, 54/4 (1964), p. 479.

9 Susie Barson, 'Infinite Variety in Brick and Stucco: 1840–1914', in Andrew Saint (ed.), *London Suburbs* (London, 1999), p. 81. For extended discussions of the interrelated development of London's rail network and railway suburbs, see Jackson, *Semi-Detached* London, chs 13 to 15, and Olsen, *Growth of Victorian* London, ch. 7.

10 Quoted in Todd Kuchta, *Semi-Detached Empire: Suburbia and the Colonization of Britain, 1880 to the Present* (Charlottesville, VA, 2010), p. 3.

11 Kuchta, *Semi-Detached Empire*, pp. 23, 85.

12 Ibid., p. 100.

13 Dyos, *Victorian Suburb*, p. 123.

14 Olsen, *Growth of Victorian London*, p. 265.

15 Howard Davis, 'Explicit Rules, Implicit Rules and Formal Variation in Vernacular Building', *Perspectives in Vernacular Architecture*, 4 (1991), pp. 53–4.

16 Ibid., p. 61.

17 Ibid., p. 63.

18 Frank H. Swinstead, 'Hornsey School of Art: Head Master's Statement of Organisation and Curriculum', 1911, in Hornsey School of Art, Minute Book of the Committee Meetings of the Hornsey School of Art, unpublished, 1901–15, n.p.

19 *North Middlesex Chronicle*, 'Hornsey Government School of Art', clipping pasted into the Minute Book of the Committee Meetings of the Hornsey School of Art, 1882–1901, between entries for 30 October 1884 and 8 December 1884, unpublished, n.p.

20 R.G. Lee (Honorary Secretary, Hornsey School of Art), letter to the Education Committee, Middlesex County Council, 11 February 1907, unpublished, n.p.

21 Battersea Polytechnic Institute, *Prospectus of Day and Evening Classes, Session 1894–5* (London, 1894), p. 22.

22 Ibid., pp. 8–10, 12, 15.

23 Ibid., p. 10.

24 Battersea Polytechnic Institute, *Prospectus of Day and Evening Classes, Session 1895–6* (London, 1895), p. 7.

25 Battersea Polytechnic Institute, *Prospectus of Day and Evening Classes, Session 1896–7* (London, 1896), p. 9.

26 Battersea Polytechnic, *Prospectus of Day and Evening Classes, Session 1911–12* (London, 1911), p. 171.

27 Battersea Polytechnic Institute, *Prospectus 1896–7*, p. 14.

28 Battersea Polytechnic, *Prospectus of Day and Evening Classes, Session 1898–9* (London, 1898), p. 49.

29 Battersea Polytechnic, *Prospectus of Day and Evening Classes, Session 1903–4* (London, 1903), p. 7.

30 Battersea Polytechnic, *Prospectus of Day and Evening Classes, Session 1904–5* (London, 1904), pp. 20–21; Battersea Polytechnic, *Prospectus of Day and Evening Classes, Session 1906–7* (London, 1906), p. 19.

31 Battersea Polytechnic, *Prospectus of Day and Evening Classes, Session 1909–10* (London, 1909), p. 188.

32 During this period, the architectural profession in Britain was undergoing a gradual process of professionalization. The principal system of architectural education had previously been articled pupillage, with students advancing large premiums to private architects in return for training, meaning that the profession was effectively closed to the lower classes. The first university architecture school was established at University College London in 1841. In 1862, the Royal Institute of British Architects (RIBA) introduced a voluntary professional examination in response to petitioning from students, and in 1882 the examination was made compulsory. RIBA developed a system of official recognition for architecture schools with the establishment of its Board of Architectural Education in 1904. Courses such as those at Battersea therefore offered working-class students a viable route into the profession for the first time. For detailed discussion of the development of architecture as an academic discipline, see Mark Crinson and Jules Lubbock, *Architecture: Art or Profession? Three Hundred Years of Architectural Education in Britain* (Manchester, 1994).

33 Camberwell School of Arts and Crafts, *Prospectus and Timetable* (London, 1898), p. 3.

34 Ibid., p. 7.

35 Camberwell School of Arts and Crafts, *Prospectus and Timetable* (London, 1900), p. 7.

36 Camberwell School of Arts and Crafts, Joint Advisory Sub-Committee minutes, 7 May 1900, unpublished, n.p.

37 Camberwell School of Arts and Crafts, *Prospectus and Timetable*, 1903, p. 4.

38 Camberwell School of Arts and Crafts, 1898, n.p.

39 Ibid., p. 7.

40 Battersea Polytechnic Institute, 1894, p. 8.

41 Battersea Polytechnic, 1898, p. 49.

42 Battersea Polytechnic Institute, 1894, p. 11.

43 Battersea Polytechnic Institute, 1896, p. 14; Battersea Polytechnic, 1903, p. 7.

44 Battersea Polytechnic, 1903, p. 12.

45 Hornsey School of Art, Minute Book of the Committee Meetings of the Hornsey School of Art, 1901–15, n.p.

46 Hornsey School of Art, *Twenty-Fourth Annual Report*, 1906, n.p.

47 We might suppose that the furthest outlying students, most notably one in Bromley in south-east London, had moved house during the course of their studies. Discounting these, the average distance travelled to college by each student was around 2 km.

48 Jackson, *Semi-Detached London*, p. 169.

49 Crowther, 'Et in suburbia ego', p. 154.

50 Mary Corbin Sies, 'Towards a Performance Theory of the Suburban Ideal, 1877–1917', *Perspectives in Vernacular Architecture*, 4 (1991), pp. 197–9.

51 Ibid., p. 206.

52 Dolores Hayden, *Building Suburbia: Green Fields and Urban Growth, 1820–2000* (New York, 2003), p. 45.

53 Ibid.

54 Robert M. Fogelson, *Bourgeois Nightmares: Suburbia, 1870–1930* (New Haven, 2005), p. 56.

55 Olsen, *The Growth of Victorian London*, p. 18.

56 Shales, 'Constructing the Empire State Building', p. 141.

57 *News of the World*, 'London Municipalities', 14 October 1900, p. 3.

58 Camberwell School of Arts and Crafts, Joint Advisory Sub-Committee minutes, 1900, n.p.

59 Steve Pile, *Real Cities: Modernity, Space and the Phantasmagorias of City Life* (London, 2005), p. 180.

60 Ibid., p.1, original emphasis.

61 Ibid., pp. 167–9.

62 Ibid., p. 176.

63 Ibid., p. 173.

64 Ibid., p. 6.

65 Ibid., pp. 11–12.

66 Ged Pope, 'Deep in South London', *Literary London: Interdisciplinary Studies in the Representation of London*, March 2008.

67 Iain Sinclair, *London Orbital* (London, 2003), p. 158.

10

The Guild of All Arts in Scarborough, Ontario: Craft, Community and Memories

Alla Myzelev

Introduction

What constitutes a craft community? Often it is not a nation, country, town but smaller gatherings of individuals who work together to create and disseminate crafts. Academic understandings of such terms as community and village (see Cooke, Chapter 1 in this volume) have become more nuanced and complex to include not only ways of life and social relations, but also the production of material culture. This chapter looks at how material culture, namely handicrafts, influenced day-to-day existence, strategic planning, and the formation of communal and personal identities using the example of the craftmaking community and hotel named the Guild of All Arts (1932–86). Throughout its existence this Canadian Toronto-based craft enterprise negotiated between tradition and innovation in craftmaking, and between individual creativity and collective identity as a community. Commerce and nostalgia were combined in the Guild of All Arts (GAA) to maintain a handmade economy and to produce closer-knit associations between makers. As numerous scholars have demonstrated,[1] nostalgia for a romanticized past communal life is as much a product of modernity as any technological innovation. The desire to create a viable community that could sustain itself and at the same time produce distinctive art and craft objects was present in Toronto from the early twentieth century. The idea of an old-fashioned Guild combined with expert management contributed to GAA's longevity and success. This essay investigates how a sense of nostalgia for an idealized past and a striving for communal living shaped the Guild of All Arts. Located in the Scarborough area in what is now eastern Toronto,[2] the GAA was inspired by the British Arts and Crafts movement[3] and Guild Socialism.[4] It managed to avoid becoming a charitable organization and gave significant impetus to the development of studio crafts in Ontario during the Depression years and afterwards.[5]

The GAA was founded in 1932 by Herbert Spencer Clark (1903–86) and Rosa M. Hewetson Clark (1888–1981), both of whom were interested in socialist theories, in particular in the Guild Socialism advocated by Robert Owen. Before their marriage in 1932 Herbert Clark made an extensive tour of Europe and the Soviet Union (1931), studied social and economic management and visited numerous communities and cooperative factories and associations. Rosa Clark, who had experience managing Hewetson Shoes Company after the death of her first husband, was familiar with Fordism, and came to reject what she saw as its exploitation of human labour.[6] After their marriage the couple became founding members of the Robert Owen Foundation, an organization that promoted and financed cooperative ventures in Canada and the United States and emphasized a progressive idea of a profit-sharing model between organizers, managers and employees, based on the tenets of the nineteenth-century British socialist Robert Owen.[7] In addition, the Robert Owen Foundation followed the principles of Guild Socialism which, in turn, was inspired by British Arts and Crafts movement. The ideas of Guild Socialism included equality, happiness through manual labour, and workers' control of industrial development through the rejection of large factories and the founding of small and medium-sized workshops. Slowly, the Clarks moved away from the doctrines of Guild Socialism toward more moderate socialist positions, as their experience showed them the importance of organizational leadership for any enterprise, including craft communities such as the GAA.[8]

The case study of the GAA not only complicates further the binary of modern and antimodern,[9] but also explains how a longing to recreate an idyllic cooperative past can turn into a positive creative force. This chapter will explain various functions of the GAA, including the revitalization of textile production techniques and the protection of architectural structures slated for demolition, concentrating on two of the most significant aspects of the GAA. After providing an overview of the enterprise's history and operations, the discussion will address its promotion of weaving on portable looms. This weaving revival influenced Canadian craft history and practice in profound ways. My investigation will then turn to the GAA's collective efforts to preserve parts of the buildings that, over several decades, were destroyed in Toronto. This eclectic collection of architectural remains became one of the most long-lasting traces of the community's existence.

Guild of All Arts: Historical Overview

The late 1920s was an exciting and vibrant time for Toronto: at over half a million strong, Canada's second-largest city was an important railway hub and a significant financial centre second only to Montreal in Quebec. During this period the city's design finally shook off the historicism and 'patina' of various romantic turns of nineteenth-century architecture. Designers became more preoccupied with modern articulations of Art Deco principles that included a restrained application of decoration, an emphasis on straight and angular lines, and the utilization of taller office buildings featuring distinct set-backs and narrowing lines. When the Depression hit Toronto after New York's stock exchange crash in 1929, most of the booming manufacturing and financial developments came to an end. The hardships of the economic decline forced some people to alternatives, including crafts and producing goods by hand. Several contemporary newspapers, for instance, featured articles on men and women who, facing unemployment, turned to practising crafts in order to make a living. One such article considered an unemployed young man who learned to make ceramic pots. Interestingly, these reports rarely made mention of how viable or profitable these new occupations were.[10] At the same time, the shortage of jobs and finances also convinced some of the importance of communal living and shared social responsibilities. Among those who considered these ideas was Rosa Clark, who, in 1932, lamented that the Great Depression had hit the Canadian province of Ontario very hard and she looked for solutions in past social experiments, such as those attempted by Robert Owen.[11] Her answer was the founding, together with her husband, of the GAA (Figure 10.1).

Numerous craftspeople came to live in Guildwood, as it became known, by invitation of the founders, to produce objects that were later sold in two craft outlets: one shop on the property of the Guild and the other on Bloor Street in downtown Toronto. Soon the GAA developed into a vibrant and prosperous community that encompassed several craft departments, including metalsmithing, sculpture, woodcarving, ceramics, weaving and furniture-making. On offer were various handcrafted objects, including textiles, woodcarvings, ceramics and stone sculptures, all of which showcased Canadian material and local flora, fauna and historical events such as contact between aboriginal cultures and European settlers, and the unity of provinces under Confederation. The stores served not only as a means of fundraising but also aimed to show the diversity of Canadian handicrafts. The GAA shops also sold items distributed by the Canadian Handicrafts Guild[12] and by other Toronto-based craft workshops. The Clarks were interested in Aboriginal culture, as well, but as with many of their contemporaries in the 1930s and 1940s they believed that Native culture and handicrafts were slowly disappearing.[13] However, led by the Clarks, the GAA took special care to find and sell objects that were produced by Aboriginal people.[14] Judging from the variety of items for sale, the couple wanted to emphasize local, regional and national

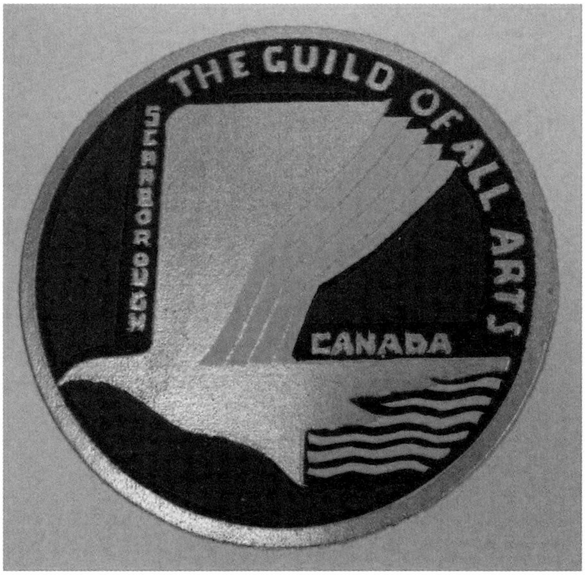

10.1 Logotype of the Guild of All Arts, c. 1931; designed by Fred Rowley, 1933. Made as a stained-glass piece in 1937 by C.J. Travers.

handicrafts. As the retail outlets they served were tourist destinations, the GAA was interested in raising awareness of local products and disseminating an eclectic version of the Canadian cultural identity.

The GAA was also strongly committed to teaching and raising the profile of handicrafts and applied arts. Experienced craftspeople taught apprentices in the workshops, either to work eventually side by side with more experienced people in the Guild or to establish businesses of their own. Several important craftspeople were educated in the Guild and then opened their own studios after economic hardships had lessened.[15]

In general, interest in alleviating the poverty of craftspeople during the Depression, disseminating craft-related knowledge, and creating a cooperative mode of production and consumption, as well as striving for self-sufficiency, positioned the Guild within the framework of the Arts and Crafts movement. At the same time, Rosa and Spencer Clark were keenly aware of the need to transform the profile of crafts from traditional objects to stylistically innovative design products that attracted discerning consumers. The Guild's close location to Toronto and the workshops on the premises offered Torontonians an enjoyable day trip. The workshops were open to the public and visitors were often able to see craftspeople at work or view a specific craft demonstration. Capitalizing on the interest of the general public, the management decided first to offer meals and then, later, overnight accommodation in what became known as the Guild Inn, a large sprawling mansion that was initially the residence of the craftspeople. The Clarks, who were both managers and curators of the Guild, strove for the enterprise to have as little financial dependency as possible and applied their business acumen to this end. To make room for hotel guests, craftspeople and staff moved to other smaller buildings.[16] The hotel soon became the main hub of the small village that the GAA had become. It included rooms designed in various ways, ranging from Edwardian eclectic interiors to Art Deco streamlined decor. All the furniture and decor elements were made on the premises and if a visitor wanted to commission a similar piece he or she was able to do so. In addition, the communal areas of the hotel were decorated in a distinctively diverse Edwardian style featuring antiques and objects produced by the GAA's craftspeople. Some of the decor items even had tags directing visitors to the gift shop where they could purchase identical or similar items. The gift shop offered not only works from the GAA craftspeople, but also objects produced by other local makers and those distributed through the Canadian Handicrafts Guild. The GAA had taken objects from Toronto craftspeople who were willing to sell through the outlet and pay modest commission. One of the most important areas of the GAA was textile-making. Weavers produced bed linen, including bedspreads, on the premises and these household textiles were offered for sale in the shop. In general, textile production was one of the significant contributions of the Guild to craft education.

Weaving Revival

The early years of the GAA's existence coincided with the further incorporation and adaptation of craft processes by factory-based industries. Tanya Harrod has observed that in Britain during the 1930s the craft techniques of fabric-dyeing, for example, were appropriated by large corporations, although the fact that the technologies, patterns and skills were actually taken from craftspeople was rarely recognized or acknowledged.[17] The craft of weaving was a particularly important influence on modern industrial production. Due to the limited budget of the Depression-stricken middle classes, many fashionable outfits were made of rough, homespun-looking fabrics. Tweeds were particularly popular for both men's and women's coats and jackets, whether or not these were hand-made. Practical and durable fabrics were also fast becoming an important part not only of fashionable clothing but also of interior decor. British textile artist Ethel Mairet argued in her book *Hand-Weaving To-day* (1939) that hand-weaving can contribute tremendously to the interior design of modern buildings.[18] She conceived of modern textiles in the style of the Bauhaus produced in small workshops such as the 'Welsh woolen mills' or by French artisan weavers.[19] Mairet's major contribution was her belief in collaboration between small community-based workshops and larger industries.

The Clarks had a similar approach with their weaving department. Rather than concentrate on revitalizing particular traditional motifs or trends, they advocated the use of handlooms for creating both contemporary and traditional textiles. The Guild's woodworking department gained particular recognition for its creation of the light, cheap and easy-to-operate Guildcraft Thackeray Looms and the Guildcraft Silent Speed Looms (Figure 10.2).[20]

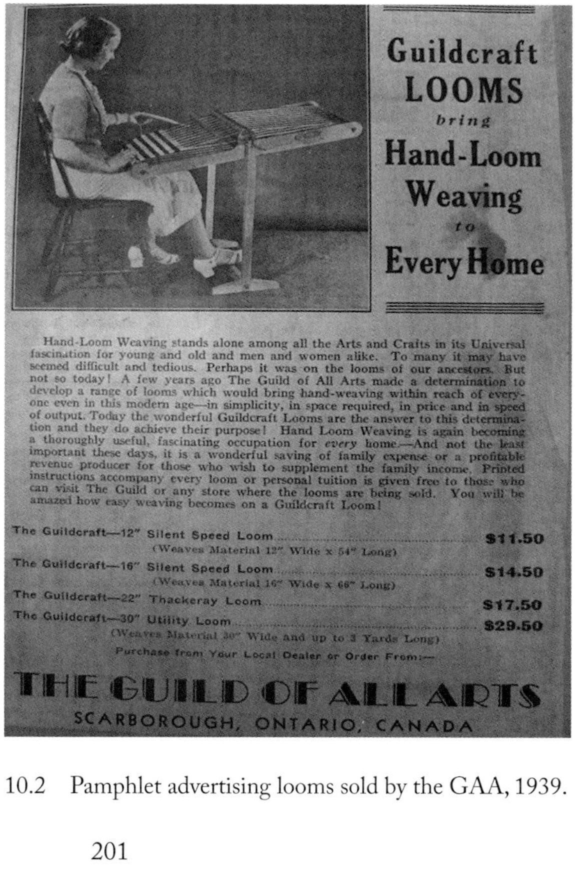

10.2 Pamphlet advertising looms sold by the GAA, 1939.

The fact that the woodworking department was working together with weavers to satisfy the demand for looms shows the importance of weaving in the economic and educational outlook of the GAA and more widely. The Clarks based their loom models on a method which was developed by Anne Thackeray, who, together with her late husband, a noted member of the English Arts and Crafts movement, developed the loom. On a visit to England in 1936, the Clarkes studied with Thackeray. The main difference between the loom models that the Guild produced and used and other models was their size and portability. Both models were much smaller than those typically produced and thus could be used anywhere. The booklets accompanying the looms claimed that learning to weave was very simple:

> the Thackeray Patent portable loom will weave an unlimited variety of materials and patterns … Those wishing to learn the art of weaving to profitably produce woven articles for use or for sale will find the Thackeray Loom most adaptable for their needs. [21]

The objects that were produced demonstrate at least one important aspect of craft production, namely its collaborative and communal nature. By providing conditions ripe for collaborations between the woodworkers and metalsmiths making the looms, and the designers and weavers producing cloth, the GAA differentiated itself from many other craft studios that concentrated on one particular type of production. This collaboration and cooperation of different craftspeople, artists and visitors helped to create a more communal feeling.

During the late 1930s and 1940s, weaving became the main project of the Guild, although other departments such as metalsmithing and leather-making were also successful. While all media branches offered courses and workshops for the general public, over the years weaving on the Thackeray loom became the main educational undertaking of the Guild. The Clarks were keen on teaching people to weave because of what they perceived as both moral and economic benefits. Part of this was obviously the desire to sell the looms, which were offered at $17.50 per Thackeray loom and $11.50 to $14.50 for the Silent Speed loom. At one point in the late 1930s sales of looms represented over half of the revenue of the crafts in the Guild.[22] In total, about 30,000 looms built by the Guild were sold all across the world as far away as Japan. Notably, Henry Ford ordered two looms in the late 1930s for his personal use.

During the Depression years, weaving along with sewing and knitting became particularly important skills to help women create and maintain their identity as fashionable and therefore modern.[23] Making one's own fabric could also be valuable if one wanted to keep up with fashion, especially because most women attempted to create clothing as cheaply as possible to leave more of their income to purchase the necessary footwear that was impossible to produce at home. The Clarks attempted to enlist the help of two large department stores in Canada: the Eaton Store and the Simpson Company. Their support was important for two reasons: first, the stores afforded space and a central location in Toronto. For example, the downtown College Street Eaton's store was one of the most popular shopping facilities in mid-century Toronto. The Clarks also expressed the hope that they would sell the hand-woven fabric in other Eaton Stores across the country. Second, and more important, the two department stores were influential tastemakers for middle- and lower-middle-class Canadians. For example, Eaton's display of contemporary furniture in Art Deco style in the 1930s became one of the primary reasons that the style gained general popularity in the country.[24]

The department stores then jumped on the bandwagon of fabric production and promoted new types of thread and yarn for weaving and knitting, along with a variety of fabrics. The GAA fed into this profitable market, which was used by the Clarks in several different ways, through production of looms, offering weaving lessons and selling fabric made in the GAA's workshops. The manufacture and sale of looms was the most important because it brought revenue and ensured that weavers came back for lessons. This arrangement also benefited Eaton's store because it attracted new clientele from the lower middle classes who normally could not afford anything in the store but potentially could learn weaving and buy thread for their projects. Classes were offered on the premises of the GAA, in the Eaton Store, in the Guild's downtown shop on Bloor Street and occasionally in other venues.[25] In addition the GAA attempted to create a hierarchy between amateur weavers to whom they sold the looms and professional ones such as Grace Poque, a master weaver who worked with the Clarks in the 1930s and 1940s.

During the late 1930s Spencer Clark also organized a weaving association, the Ontario Association of Hand-Loom Weavers, for which Toronto-based furniture company Ridpaths Limited provided financial backing.[26] The Simpson Company was not interested in supporting the Association, which at that point consisted of the Clark family and GAA employees. This institutionalization and professionalization of weaving paralleled the creation of various craft associations such as the Canadian Handicrafts Guild that took on the promotion, education and setting of professional standards. Simpson's did not collaborate with the GAA, but they did open a handicrafts school adjacent to the needlework department in their department store, albeit only for a short period. Both Simpson's and Eaton's stores featured occasional demonstrations of weaving and knitting techniques. Both were sold materials for weaving and knitting and therefore both promoted handicrafts by organizing demonstration of new fashions that could be made with the items purchased from the department stores. There were many advertisements and frequent reports in the newspapers about new weaving techniques. As one writer put it:

> Weaving is the newest and rapidly becoming the greatest vogue for all knitters and fancy-workers. It is incredibly fast on the new Guildcraft-Thackeray Looms. Endless new designs, effects and materials become possible. It's fascinating fun and highly profitable.[27]

Roberta Noble, the instructor from the Guild's Bloor shop, was praised as one of the most knowledgeable and most industrious weavers.[28] An image accompanying an advertisement shows a young woman, neatly and fashionably dressed, coiffed and smiling, obviously enjoying the process of weaving. Some newspapers commended weaving as useful and profitable for both men and women and, indeed, many photographs of the GAA feature men weaving.[29] The Clarks believed that weaving would help both sexes to become more industrious and aid them financially by sparing them from buying fabric and/or ready-made wear.[30] For example they organized weaving demonstrations for 150 Ontario teachers[31] in the summer of 1935 and explained the benefits of teaching weaving to school children. The GAA also donated a few looms to the local Scarborough school with the provision that the students, girls in particular, would learn the craft in the classroom.

The Clark family lived their beliefs. Rosa Clark and her three daughters, Rosemary, Ruth and Dorothy, wove fabric for their own clothes (Figure 10.3).

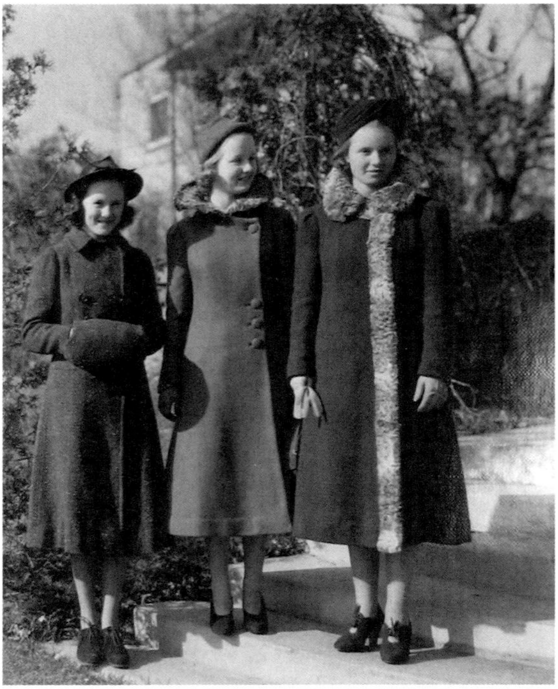

10.3 Rosa Hewetson Clark's daughters demonstrate the coats made from hand-woven fabric.

An article in the *Globe & Mail* (2 February 1938) mentioned that in one day Rosa had created enough cloth for a new skirt. In the morning she began weaving a piece of bright plaid material; by noon the fabric was finished. For buttons she asked artisans in the workshop to create circles, which were hammered out of pewter. In their spare hours, over a period of four days, Rosemary, Ruth and Dorothy, on a 30-inch utility loom, wove 7.5 yards of fabric for each of their winter coats. They modelled the fabric for these and for display in Eaton's store window. The fact that the young Clark women created coats for themselves advertised their industriousness and domestic virtues. The image published in the newspaper, emphasizing the youth of these young women and their wonderful smiles, also stresses the modernity, the novelty of the coat designs. There was nothing stereotypically *homespun* about the clothing presented. The tweeds and other fabrics produced on the looms were then made into very up-to-date garments; one might even say they reflected cutting-edge designs that were marketed through advertising in the shops and the display windows at Eaton's store. The jackets were sophisticated and urban, echoing the trends of the season. The clothing appealed to a middle-class clientele because it was fashionable, contemporary and urban, very similar in style and cut to what was sold at the time in Eaton's and Simpson's department stores.[32]

In cooperation with Monarch Knitting, the 100-inch loom, the largest of its kind, was displayed in 1935 at the Canadian National Exhibition (CNE) and later in the Guild's own booth. The speed and beauty of the weaving was also recognized at the annual CNE in August 1940 when 15-year-old Rosemary won a competition for hand-woven Army, Navy and Air Force scarves. She was the fastest weaver, producing a scarf in two hours.[33] Grace Poque, the wife of the Guild Inn manager, was also a prize-winning weaver.[34] At the CNE in August 1941, in her second year of training, she won eight prizes out of nine entries. The previous year she won six prizes.[35] This showcase of professional weaving at the Toronto-based national exhibition demonstrated the high professionalism of the Guild's weavers and helped to promote this activity as patriotic and useful for the home as well as for soldiers at the front.

Weaving is perhaps the most dramatically performative of all the textile arts, with the physical exertion required to prepare the heddles, the broad arm movements required to throw the shuttle to lay the weft and beat in the weft against the fell of the cloth when the reed is pulled by the weaver. It demands the coordinated movement of the hands and feet. Its consistent rhythm was believed to be calming and therapeutic. The weaver concentrates on the movement and the motif and, as a result, the final object and the purpose of making is realized. Through the weaving performance (for the audience or the weaver) the craftsperson is detached from the final product and dedicates his or her effort to the process itself. Thus Rosa Clark and her fellow craftspeople employed the performative and adaptable qualities of hand-loom weaving. Rather than follow traditional crafts and cultivate old-fashioned motives, they concentrated on the process of weaving and the joy of making fashionable items in financially desperate times. Along with her daughters, Rosa Clark challenged the limitations of cloth production and reformulated weaving as a modern craft. She changed the associations of weaving from grandmother's necessity to a contemporary, creative pursuit. The GAA looked at this craft as an opportunity to create something new, a rare combination of handicrafts and fashion.

Community in Flux: From Craft to Architecture

The success of the weaving department continued far beyond the Depression years. Except during the Second World War when the GAA was taken over by the federal government (it became a site for military training and rehabilitation hospital for

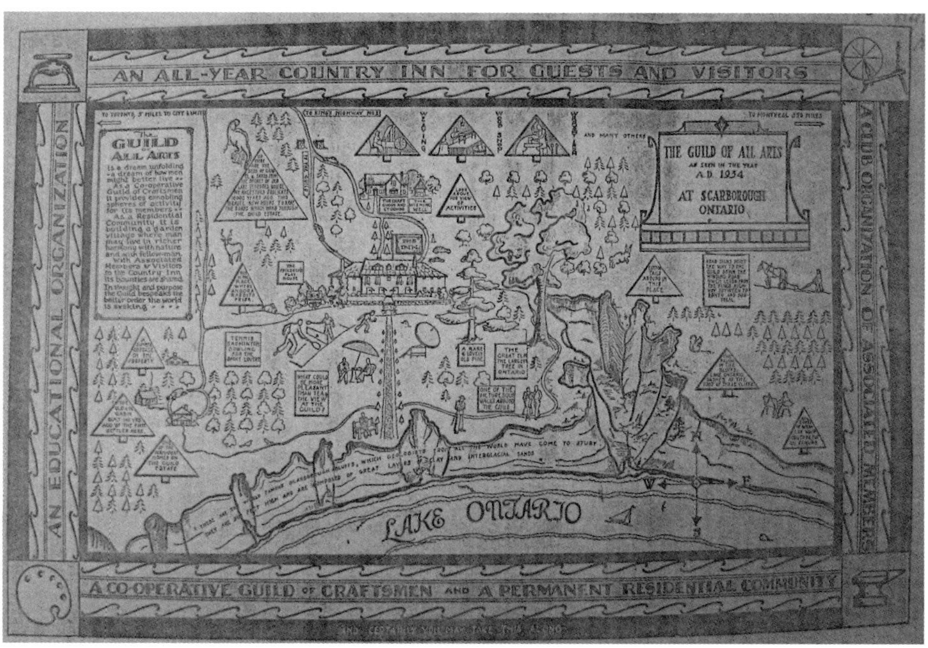

10.4 Map of the Guild of All Arts given out to visitors between 1939 and 1949.

soldiers), weaving lessons were offered on the Guild premises and in schools across the city throughout the 1940s and 1950s.[36] However, the orientation of the GAA had started to change in 1950s mainly due to the changing economic circumstances of many craftspeople. By the 1950s, in spite of a lively resort life at the weekends, vibrant craft production and a multitude of cultural events,[37] the Guild found it difficult to attract experienced craftspeople and apprentices (Figure 10.4).

Many of the experienced craftspeople left the GAA to open their own private studios or work elsewhere. The low wages and communal living arrangements proved to be challenging for many. While during the 1950s craftmaking and especially weaving lessons remained an important revenue source for the GAA, the booming area of the business in these years was the Inn. The resort, situated above Lake Ontario, although it did not have direct access to the beach, provided picturesque views and an easy escape from rapidly urbanizing and industrializing Toronto. Towards the early 1960s The Guild Inn also had to adapt to a big shift in the local patterns of travel and leisure since few people wanted to stay there overnight. For most Torontonians, and tourists from outside of the city who preferred to stay downtown, the GAA became a day-trip destination rather than overnight one.

To address the changes that were happening around them, starting from 1950s, the Clarks became more interested in the preservation of the architectural heritage of Toronto. The building boom and prospering economy after the Second World War brought rapid changes in the built environments of the urban centres in North America. Many Neoclassical and Art Deco structures were demolished to make way for functional modernist edifices. The Clarks, especially Herbert, became very involved with the preservation efforts around Toronto architecture. By the late 1970s the idea of preserving and commemorating the past far outweighed their desire to practise and disseminate craft knowledge or other communal artistic undertakings. The Clarks concentrated their attention on running the hotel and providing entertainment for their guests. Thus, slowly the GAA became more of a local attraction and historical landmark than an active craft community. In light of the dearth of experienced craftspeople and the lack of overnight tourists, Clark, an experienced businessman, realized that holding on to almost 500 acres of land was also not important to his present-day objectives.

Additional hardships and financial liabilities appeared in 1967, when Scarborough was merged with Toronto. As a result, property taxes on Guild land skyrocketed and the community declined further. The Clarks were forced to sell all but 50 acres of the Guild Inn Resort. Therefore they focused on other options, founding Guildwood Development and hiring one of the most reputable town planners in Ontario, Eugene G. Fluid,[38] to create a community rooted in the Garden Suburb movement of the turn of the century.[39] Guildwood Village was laid out with winding roads and shaded pavements.[40] Houses were reasonably priced ($25,000 to $30,000) and came in 12 designs to avoid monotony and provide choice. This exemplary community catered to 'imagined' middle-class families and followed the belief that a well-designed community and adequate housing can guarantee moral respectability.[41] The community even had symbolic gates, reminding one of gated communities of a later period and suggesting an increased sense of security. The Clarks transported the iron gates from a demolished barracks building in Toronto. These grand cast-iron gates, which dated from 1839, seemed somewhat out of place in a 1950s residential development for the middle and higher middle classes, yet together with the coat of arms designed for the village by craftsmen of the GAA, it was in line with the Clarks' motto chosen for this project: 'Let us mingle the beautiful with the useful'.[42] The old-historical gates symbolized the threshold between the community of this housing estate based on the most up-to-date theories of house planning and the rest of the suburban Toronto. It also testified that Guildwood Village was not a part of the GAA and therefore helped to create a very clear divide between the two projects. Finally, it emphasized the exclusivity of the community as a gated community for the middle class, something that existed in Toronto as early as the end of the nineteenth century.[43] Guildwood Village then continued the tradition of communal enclaves such as Rolling Hills and Hidden Hills in the USA and Wychwood Park in Toronto.[44] These communities were structured around the aesthetic sensibilities and common social and economic classes of their members. The fact that Guildwood Village was destined to turn into a middle-class upscale neighbourhood is somewhat surprising given the charitable and socialist inclinations of the Clarks. Yet it shows that the survival of the GAA itself took priority over adherence to long-held social ideals. As Spencer Clark's papers demonstrate, he was now less preoccupied with Guild Socialism and the ideals of labour and equality after the Second World War. As will be shown below, architecture and architectural preservation became much more important to the Clarks at this time.

While Guildwood Village was growing and becoming more a part of the urban fabric of Toronto, the GAA resembled a distinct community less and less. This brought an increasing sense of disappointment to the Clarks. Finally, in 1978 the Clarks decided to sell the Inn to the city on terms beneficial for the City of Toronto. The City of Toronto was guaranteed against losses, and the Clarks remained in charge of the Inn for five more years to facilitate the transition. They would not take a salary but returned any profit to the city treasury. In exchange, Spencer Clark demanded recognition of the entire area of the GAA (not including Guildwood Village) as a museum or heritage site (Figure 10.5).

Clark insisted on this because he feared that his collection of salvaged architectural details would be disposed of after he retired from the GAA. The Clarks started collecting the architectural details around the time they founded the GAA, reflecting a long-term fascination with these materials, but it was not until the 1950s and 1960s, when Toronto was swept by rapid industrialization and modernization, that they became really concerned about the disappearance of notable early twentieth-century buildings.[45] Clark was often alerted by his friends and acquaintances that an important building was being demolished; he would then try to salvage what he could. For instance, prompted by a well-known Toronto architect, Eric Arthur, Clark managed to transport a set of four Ionic columns in 1973 from the Neoclassical Banker Bond Building (built in 1920 by J.A. Mackenzie) to the Guild premises. The Clarks were interested in several different architectural styles; for example, in 1972, they salvaged two Art Nouveau panels made of Ohio Limestone. These were taken from the Canadian Bank of Commerce building designed by the architectural firm of Frank Darling and John A. Pearson in 1899. Varying in size and style, the architectural details were then

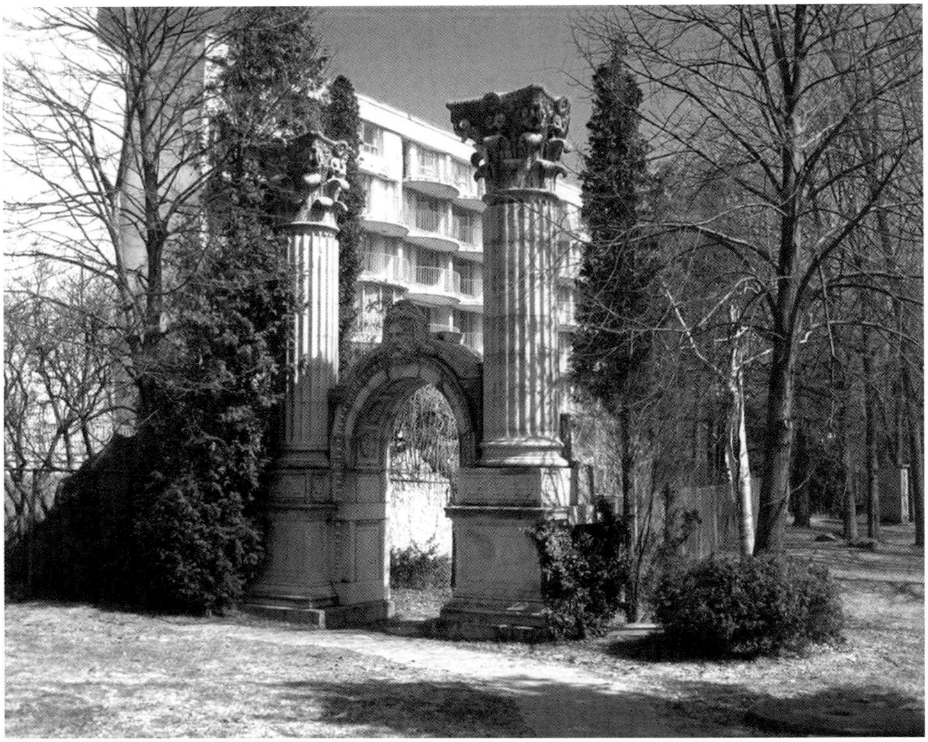

10.5 Entranceway from Toronto Dominion Bank Building. Architect Ronald Thom. Built in 1912, demolished in 1966.

moved to the resort and installed there to complement visitors' experiences in the park. In 1978, *The Star* (Toronto) newspaper profiled six new bas-reliefs that were salvaged from a Bank of Montreal building. It took Clark five years of negotiating to receive these sculptures, created by reputed Canadian sculptors Emanuel Hahn, Elizabeth Wyn Wood, Frances Loring and Florence Wyle, among others.

Clark approached this undertaking as a serious collecting enterprise. He even appointed Hugh Walker, an amateur architectural historian, as curator of the collection and published a catalogue in 1982. Seen against the rapid changes that were taking place in Toronto, Clark's interest in older architecture coincided with strong opposition to the demolition of many buildings, which architects such as Eric Arthur organized in 1950s and 1960s.[46] Perhaps, therefore, the most appropriate question to ask is: how significant was the GAA's desire to preserve and revive crafts and salvage architecture heritage?

Adapting to economic fluctuations such as the Great Depression, the Second World War and 1950s and 1960s development, the Guild was transformed by the 1980s into a monument to Toronto's architectural history. Currently on top of the Scarborough Bluffs, in the eastern end of the City of Toronto, overlooking Lake Ontario, it is open to visitors as a park. The hotel is closed and boarded up; most workshops are also locked; while remnants of architectural structures have been left unattended and unmaintained.

The City of Toronto is now attempting to find a private corporation to take over the hotel and adjacent property.[47]

Conclusion: From Nostalgia to … Nostalgia

According to a 1949 advertisement, 'The Guild Inn is many things. It is a treasure land of early Canadiana. It is a blend of Old World character with the exciting new'.[48] Between its inception in the 1930s and its demise in the 1980s, the GAA changed its goals from being a community geared towards preserving, disseminating and practising crafts and artistic collective living, to a community that preserved architectural remnants and crafts as they were practised from the Guild's inception to about the 1960s. The founding of the GAA, Guildwood Village and the architectural collection testify to the desire to preserve the past and make it useful to current lifestyles. The Guild had become a commemoration site of Toronto architecture and of its own past. Based on Arts and Crafts ideology, it hoped to bring together all the arts and crafts to create aesthetic living. Similar to many undertakings that adhered to Arts and Crafts ideas, not all the aspects of the GAA were completely democratic. Guildwood Village, for example, catered to well-to-do Torontonians and thus excluded less privileged home-seekers; but the craft workshops at the GAA were also allowed to continue operation for a time.

Somewhat paradoxically, the GAA, between its inception in 1932 and its closure in 1982, managed to serve as several different entities related to aesthetic and functional crafts: it was a craft community with workshops opened to the public and artists living on the premises; a profitable hotel that offered some of its fixtures for sale; the headquarters for an innovative weaving method using portable looms; and, finally, a repository for the architectural remains from Toronto's two building booms of the 1890s and the 1920s. The enterprise had started as a community and ended as an aspiring heritage site.[49] The longevity and importance of the GAA for material culture studies lies in the fact that the Clarks and their associates managed to reinvent the community a number of times to remain a viable craft centre; they put forward their belief in the preservation of the past for future generations, albeit in altered states. Ruins of the buildings became part of the garden landscape while weaving looms produced fashionable materials and metalsmiths worked on creating stylish coat buttons. At the core of the Clarks' concerns was the need to keep the enterprise going and make sure that those who were in their employ had been provided with work and income.[50]

Today, the GAA territory can be described as a cemetery of forgotten buildings. Similar to human cemeteries, this architectural site inspires feeling of longing and nostalgia for the past. Likewise, the objects produced by the GAA still appear in sales and auctions and are rescued and preserved in ways similar to those of Spencer Clark when he attempted to preserve the architectural ruins of early twentieth-century Toronto. The remains of the buildings evacuated from demolition sites together with remnants of the workshops remind visitors of the GAA's attempts to create a community that preserved and encouraged craft pursuits. Collected for over 50 years by Spencer and Rosa Clark, these Edwardian and Art Deco remains are a memento of the period when the Guild was active and when Toronto, although Depression-stricken, forged its way towards modern life.

Sources Cited

Anderson, Benedict, *Imagined Communities: Reflections on the Origin and Spread of Nationalism*, rev. edn (London and New York: Verso, 2006).

Austin, Linda Marilyn, *Nostalgia in Transition, 1780–1917* (Charlottesville, VA: University of Virginia Press, 2007).

Bates, Christina, 'Shop and Factory', in Alexandra Palmer (ed.), *Fashion: A Canadian Perspective* (Toronto: University of Toronto Press, 2004), pp. 113–38.

Bennett, Tony, *The Birth of the Museum: History, Theory, Politics* (London: Routledge, 1995).

Bocking, Stephen, 'Constructing Urban Expertise: Professional and Political Authority in Toronto, 1940–1970', *Journal of Urban History*, 33/2 (2006), pp. 51–76.

Bonis, Robert Raynes, *A History of Scarborough* (Scarborough, Ontario: Scarborough Public Library, 1965).

Boym, Svetlana, *The Future of Nostalgia* (New York: Basic Books, 2001).

City of Toronto, 'The Guild'. 2012. Available at: http://www.toronto.ca/culture/the_guild.htm.

Crawford, Gail, *A Fine Line: Studio Crafts in Ontario from 1930 to the Present* (Toronto: Dundurn Press, 1998).

Crook, Tom, 'Craft and the Dialogics of Modernity: The Arts and Crafts Movement in Late-Victorian and Edwardian England', *Journal of Modern Craft*, 2/1 (March 2009), pp. 17–32.

Dendy, William and William Kilbourn, *Toronto Observed: Its Architecture, Patrons, and History* (Toronto: Oxford University Press, 1986).

Diehl, Daniel and Mark P. Donnelly, *Elbert Hubbard: The Common Sense Revolutionary: A Business Vision for our Times* (London: Spiro Press, 2003).

Frost, Beryl, 'The Guild Inn; Unique Country Inn Promotes Arts and Crafts', *Hotel and Travel News*, 28/1 (1955), pp. 14–16.

Glass, S.T., *The Responsible Society: The Ideas of the English Guild Socialist* (London: Longmans, 1966).

Goheen, Peter G., 'The Assertion of Middle-class Claims to Public Space in late Victorian Toronto', *Journal of Historical Geography*, 29/1 (January 2003), pp. 73–92.

Grant, Jill. 'Planning Responses to Gated Communities in Canada', *Housing Studies*, 20/2 (2005), pp. 273–85.

'The Guild Inn', *Carling Hospitality*, 3/7 (April–May 1949), pp. 4–6.

Harrod, Tanya, *The Crafts in Britain in the 20th Century* (New Haven, CT: Published for the Bard Graduate Center for Studies in the Decorative Arts by Yale University Press, 1999).

Hobsbawm, Eric, *The Invention of Tradition* (Cambridge and New York: Cambridge University Press, 1983).

Howarth, Dorothy, 'First the Inn – then the Village', *The Telegram* (16 December 1961), sec. 2, n.p.

Le Goix, Renaud, 'Gated Communities: Sprawl and Social Segregation in Southern California', in Rowland Atkinson and Sarah Blandy (eds), *Gated Communities* (Oxford: Routledge, 2006), pp. 131–51.

Lidgold, Carole M., *The History of the Guild Inn* (Toronto: Brookridge, 2000).

Mairet, Ethel, *Hand-weaving To-day: Traditions and Changes* (London: Faber & Faber, 1939).

McKay, Ian, *The Quest of the Folk: Antimodernism and Cultural Selection in Twentieth-Century Nova Scotia* (Montreal: McGill-Queen's University Press, 1994).

McLeod Easton, Ellen, *In Good Hands: The Women of the Canadian Handicrafts Guild* (Montreal: McGill-Queen's University Press, 1999).

Nurse, Andrew and Raymond Benjamin Blake (eds), *The Trajectories of Rural Life: New Perspectives on Rural Canada* (Regina: Canadian Plains Research Center, 2003).

Owen, Robert, *A New View of Society and Other Writings* (London: Penguin Books, 1991).

Panayotidis, Euthalia Lisa and Paul James Stortz (eds), *Historical Identities: The Professoriate in Canada* (Toronto: University of Toronto Press, 2006).

Parkins, Wendy (ed.), *William Morris and the Art of Everyday Life* (Newcastle upon Tyne: Cambridge Scholars, 2010).

Parsons, Kermit Carlyle (ed.), *From Garden City to Green City: The Legacy of Ebenezer Howard* (Baltimore, MD and London: Johns Hopkins University Press, 2002).

Penty, Arthur Joseph, *Guilds and the Social Crisis* (London: Allen & Unwin, 1919).

Penty, Arthur Joseph, *A Guildsman's Interpretation of History* (London: Allen & Unwin, 1920).

Penty, Arthur Joseph, *Post-Industrialism* (New York: Macmillan, 1922).

Penty, Arthur Joseph, *Tradition and Modernism in Politics* (London: Sheed & Ward, 1937).

Phillips, Ruth (ed.), *Trading Identities: The Souvenir in Native North American Art from the Northeast, 1700–1900* (Seattle, WA: University of Washington Press, 1998).

Radstone, Susannah, *The Sexual Politics of Time: Confession, Nostalgia, Memory* (London: Routledge, 2007).

Santink, Joy, *Timothy Eaton and the Rise of His Department Store* (Toronto: University of Toronto Press, 1990).

Sewell, John, *The Shape of the City: Toronto Struggles with Modern Planning* (Toronto: University of Toronto Press, 1993).

Sifton, Elizabeth, 'Montreal's Fashion Mile', in Alexandra Palmer (ed.), *Fashion: A Canadian Perspective* (Toronto: University of Toronto Press, 2004), pp. 203–26.

Srigley, Katrina, 'Clothing Stories: Consumption, Identity, and Desire in Depression-Era Toronto', *Journal of Women's History*, 19/1 (Spring 2007), pp. 82–104.

Talen, Emily, *New Urbanism and American Planning: The Conflict of Cultures* (New York: Routledge, 2005).

The Arts and Crafts Society, 'Elbert Hubbard', 2011, 10 February 2012. Available at: http://www.arts-crafts.com/archive/ehubbard.shtml.

Thomson, Watson, *Pioneer in Community: Henri Lasserre's Contribution to the Fully Cooperative Society* (Toronto: Ryerson Press, 1949).

Vitale, Patrick, 'A Model Suburb for Model Suburbanites: Order, Control, and Expertise in Thorncrest Village', *Urban History Review*, 40/1 (2011), pp. 41–55.

Watson, Scott, 'Art in the Fifties: Design, Leisure, and Painting in the Age of Anxiety', in Scott Watson (ed.), *Vancouver, Art and Artists, 1931–1983* (Vancouver: Vancouver Art Gallery, 1983), pp. 89–105.

Notes

1 On creating traditions that underlie sense of community and give its members sense of common identity see Eric J. Hobsbawm, *The Invention of Tradition* (Cambridge, 1983); Benedict Anderson, *Imagined Communities: Reflections on the Origin and Spread of Nationalism* (New York and London, 2006); Susannah Radstone, *The Sexual Politics of Time: Confession, Nostalgia, Memory* (London, 2007). It is, however, very important to differentiate between community in a sense of national community and establishing national traditions, and longing for creating local and regional communities. The discussion of Guild of All Arts will deal with the creation of the local community.

2 In the 1930s Scarborough was a village outside Toronto.

3 Not only were the Clarks admirers of the British Arts and Crafts movement; they had also visited and even honeymooned in Roycroft (1895–1938), a community located in East Aurora, in Upstate New York that was built according to Arts and Crafts principles of handmade production, affordable pricing and distribution of profits among the craftspeople. Its founder Elbert Hubbard was influenced by William Morris and first set up his own private press, inspired by Morris's Kelmscott Press. Hubbard's ideas attracted a number of visiting craftspeople to East Aurora, and they formed a community of printers, furniture-makers, metalsmiths, leathersmiths and bookbinders. A quotation from John Ruskin formed the Roycroft main principle: 'A belief in working with the head, hand and heart and mixing enough play with the work so that every task is pleasurable and makes for health and happiness'. Available at: http://www.arts-crafts.com/archive/ehubbard.shtml [accessed 18 January 2012]. For more on Roycroft history and business see Daniel Diehl and Mark P. Donnelly, *Elbert Hubbard: the Common Sense Revolutionary: A Business Vision for our Times* (London, c. 2003).

4 Watson Thomson, *Pioneer in Community: Henry Lasserre's Contribution to the Fully Cooperative Society* (Toronto, 1949), pp. 44–7.

5 Throughout the nineteenth and twentieth centuries, communities of craftmakers had often encountered financial difficulties for reasons that included inability to obtain fair prices for the objects, challenges of management and organization and the overall poverty of the participants. Thus more often than not communities that were funded to help craftspeople in need ended up requiring influxes of large sums of money and essentially were not profitable business ventures. Although the GAA had encountered financial difficulties several times, in particular in the 1930s, overall it managed to avoid becoming an organization that consisted of the charity of wealthy patrons, mainly due to skilful management of the Clarks and the existence of the hotel.

This self-reliance is important not only for the history of the GAA but also for the history of the craft-related communities because it shows a unique example of self-sufficiency when sources of income such as patronage or higher prices for craft production were not available.

6 Rosa Breithaupt Hewetson, unpublished diary. This information, along with most of the other primary sources, comes from the Breithaupt Hewetson Clark Collection, a significant resource of information not only on the Guild of All Arts but also on any matter related to the lives of Rosa and Spencer Clark. The collection is located in Betty Fisher Rare Book Library at the University of Waterloo, and consists of the personal library of the Clarks, archival collections, their correspondents, promotional materials of the Guild and scrapbooks of the Clarks mainly relating to the Guild and their later project of Guildwood Village. University of Waterloo Library, Breithaupt Hewetson Clark Collection, 1.2. box 6.

7 Shortly after Clark was elected its secretary–treasurer and was responsible for the financial operation of the society.

8 Spencer Clark, 'The history of my life', unpublished biography. University of Waterloo Library, Breithaupt Hewetson Clark Collection, 1.2, box 6.

9 Some scholars see later nineteenth- and early twentieth-century interest in traditional craft and exotic culture as a desire to escape modernity and hence became a pre-modern. On anti-modernism and Canadian cultural history see, for example, Ian McKay, *The Quest of the Folk Antimodernism and Cultural Selection in Twentieth-Century Nova Scotia* (Montreal, 1994); Andrew Nurse, *The Trajectories of Rural Life: New Perspectives on Rural Canada* (Regina, 2003); Ruth Phillips, *Trading Identities: the Souvenir in Native North American Art from the Northeast, 1700–1900* (Seattle, WA, 1998); and Lisa Panayotidis, *Historical Identities: The Professoriate in Canada* (Buffalo, 2006).

10 See for instance Mrs Smith of Englehart, 'Handicrafts: A Visit to Wonderland', *The Weekly Sun*, 11 January 1934, n.p.; 'future for Young Men seen in Ceramic Work', *Globe & Mail*, 9 April 1940, n.p.

11 University of Waterloo Library, Breithaupt Hewetson Clark Collection, 1.2, box 6.

12 Initially founded in Montreal, by the early 1930s the Canadian Handicrafts Guild became the main organization that united craftspeople and those who wanted to revitalize and revive craft techniques. It promoted the handmade works through a series of exhibitions in Canada and United States. The CHG had several outlet depots in Canada and often exhibited in Toronto. For more, see Ellen McLeod Easton, *In Good Hands: The Women of the Canadian Handicraft Guild* (Montreal, 1999).

13 For instance, the Clarks' scrapbooks from 1938 to 1940 include a large article 'To the Rescue of the Indian Handicrafts' telling about objects produced by Ojibwa and other tribes and presented by the CHG exhibition in Vancouver. University of Waterloo Library, Breithaupt Hewetson Clark Collection, 1.2, box 4.

14 See for example the letter that Clark wrote to James Mitchell, probably a middleman between Aboriginal craft workshops and Guildwood. University of Waterloo Library, Breithaupt Hewetson Clark Collection, 1.2. Herbert Spencer Clark Archive-Speeches, box 3.

15 For more on studio crafts in Ontario see Gail Crawford, *A Fine Line, Studio Crafts in Ontario from 1930 to the Present* (Toronto, 1998).

16 Rosa Clark had purchased a homestead estate built in 1914 for Sergeant-General Harold C. Bickford, a retired American soldier, in 1932, shortly before the founding of the Guild and her subsequent marriage to Spencer Clark.

17 Tanya Harrod, *The Crafts in Britain in the 20th Century* (New Haven, CT, 1999), p. 121.

18 Ethel Mairet, *Hand-weaving of To-day: Traditions and Changes* (New York, 1939), pp. 28–42.

19 Ibid., 25.

20 The Guildcraft Thackeray loom uses string heddle bars (similar to using a blackstrap loom), but the brackets at the side hold the heddle sticks in the required up/down positions. Two bars fit in the lower front part of the bracket and two in the back when at rest. This type of configuration makes making a warp much easier than on larger traditional looms. The Guildcraft Silent Speed looms resemble closely the Thackeray style but are smaller and thus more portable, yet allow making narrower pieces of fabric.

21 *Learn to Weave Easily and Quickly on a Thackeray Loom*, Promotional Booklet published by the GAA, 1936.

22 Financial statement of the GAA from 1934, 1935, 1937, University of Waterloo Library, Breithaupt Hewetson Clark Collection, 1.2, box 7.

23 Katrina Srigley, 'Clothing Stories: Consumption, Identity, and Desire in Depression-Era Toronto', *Journal of Women's History*, 19/1 (Spring 2007), pp. 82–104.

24 On Eaton Store and its influence on the dissemination of innovative style in Canada see Sandra Cohen-Rose, *Northern Deco: Art Deco Architecture in Montréal* (Montreal, 1996) and Joy L. Santink, *Timothy Eaton and the Rise of his Department Store* (Toronto, 1990).

25 The Clarks were constantly attempting to expand their area of distribution and education. Clark proposed in a letter to C.L. Burton, president of Simpson Department Store Chain Company, to teach hand-weaving classes on the sixth floor of the Toronto Store in 1938. University of Waterloo Library, Breithaupt Hewetson Clark Collection, 1.2, box 4, item 92.

26 Gail. Crawford, *A Fine Line: Studio Crafts in Ontario from 1930 to the Present* (Toronto, 1998), p. 186.

27 *Learn to Weave Easily and Quickly on a Thackeray Loom*, Promotional Booklet.

28 Between 1932 and 1941 the Clarks employed numerous craftspeople. As Gail Crawford notes, it is very difficult to understand who was employed by the Guild from Spencer Clark's voluminous correspondence and whom he was hoping to 'lure' (Crawford, *A Fine Line*, pp. 185–6) to the Guild. It seemed that the main drawback for the craftspeople was the very low commissions and the lack of opportunity to sell their products at other venues if they chose to become associated with the Guild.

29 'Student Visit Guild', *Globe & Mail*, 30 July, no year. University of Waterloo Library, Breithaupt Hewetson Clark Collection, Herbert Clark Scrapbooks 1937–43.

30 Mrs Smith, 'Handicrafts', *The Weekly Sun*, 1934. University of Waterloo Library, Breithaupt Hewetson Clark Collection, Herbert Clark Scrapbooks 1934–36. Smith relates her conversation with Rosa and Spencer Clark and notes that hand-woven cloth is more economical and better suited for Canadian women.

31 'Weaving Demonstrated: 150 Teachers Hear a Talk on the Value of Craft', University of Waterloo Library, Breithaupt Hewetson Clark Collection, Herbert Clark Scrapbooks 1937–39.

32 On fashion in Canada see two papers: Christina Bates, 'Shop and Factory', and Elizabeth Sifton, 'Montreal's Fashion Mile', both in Alexandra Palmer (ed.), *Fashion: A Canadian Perspective* (Toronto, 2004), pp. 113–38 and 203–26.

33 This loom, the only one of its kind in Canada, is now displayed in the Ontario Science Centre in Toronto.

34 Cherry, 'Jaunting about the EX', unidentified newspaper clipping, University of Waterloo Library, Breithaupt Hewetson Clark Collection, Herbert Clark Scrapbooks 1937–43.

35 Weaving was believed to have rehabilitation value and was taught as an occupation for the blind through the Canadian National Institute for the Blind. The latter purchased its looms from the GAA.

36 'Weaving Demonstrated: 150 Teachers Hear Talk on Value of Craft', unidentified newspaper clipping, University of Waterloo Library, Breithaupt Hewetson Clark Collection. Herbert Clark Scrapbooks 1937–39.

37 Guild Inn and the GAA often had lectures and seminars related to artistic production and general cultural topics. For example, on 15 May 1951 Irena Gartner lectured on the music of the Renaissance and on 15 June 1952 Dale Hanover delivered a talk on Dutch painting.

38 Faludi was one of the most influential and authoritative urban planners in Toronto. He assumed the leadership role in creating Master Plans for Toronto in 1943 and Hamilton in 1947. On his work in Toronto see Stephen Bocking, 'Constructing Urban Expertise: Professional and Political Authority in Toronto, 1940–1970', *Journal of Urban history*, 33/2 (2006): pp. 51–76. See also Patrick Vitale, 'A Model Suburb for Model Suburbanites: Order, Control, and Expertise in Thorncrest Village', *Urban History Review*, 40/1 (2011), pp. 41–55.

39 The main advocate and developer of the Garden City and Garden Suburb movement was Ebenezer Howard (1850–1928), who, reacting against increased urbanization and industrialization of the larger cities in England, developed the largely utopian idea of cities that provide all the conditions for harmonious living together with nature. Although attempts at creating Garden Cities were not successful, Howard's influence proved long-lasting as many suburbs in England and North America attempted to implement the Garden City system of combining the best that country and city have to offer. For example, American architect Frederick Law Olmstead and his sons Fredrick Law Olmstead Jr and John Charles Olmstead successfully combined ideas of creating suburbs and suburban campuses surrounded by nature. Carlyle Parsons Kermit (ed.), *From Garden City to Green City: The Legacy of Ebenezer Howard* (Baltimore, MD, 2002).

40 The community prided itself on not cutting down trees but building around existing ones to preserve the local landscape. To emphasize the aesthetic appeal of the development, all electrical lines were put underground and the roots of the trees were drilled to allow for electricity to reach houses and backyards. Toronto Reference Library, Guildwood Village files, unidentified newspaper clipping.

41 Scott Watson, 'Art in the Fifties: Design, Leisure, and Painting in the Age of Anxiety', *Vancouver Art and Artists 1931–1983* (Vancouver, 1983).

42 Carole M. Lidgold, *The History of the Guild Inn* (Toronto, 2000), p. 34.

43 See, for example, Peter G. Goheen, 'The Assertion of Middle-class Claims to Public Space in late Victorian Toronto', *Journal of Historical Geography*, 29/1 (January 2003), pp. 73–92; and Jill Grant, 'Planning Responses to Gated Communities in Canada', *Housing Studies*, 20/2 (2005), pp. 273–85. Both note the gated communities that appeared in two affluent areas of Toronto: Rosedale and Lawrence Park.

44 Renaud Le Goix, 'Gated Communities: Sprawl and Social Segregation in Southern California', in Rowland Atkinson and Sarah Blandy (eds), *Gated Communities* (Oxford, 2006), pp. 131–51.

45 Toronto became one of the centres of international modernism from the 1950s. For example, Toronto Dominion Centre was designed by Mies van der Rohe between 1963 and 1967. William Dendy and William Kilbourn, *Toronto Observed: Its Architecture, Patrons, and History* (Toronto, 1986).

46 One example of the most heated preservationist debates was City Hall's decision to demolish Toronto Chinatown because allegedly it presented a threat to public health. The city went back on its decision after Chinatown residents and some planners and architects fiercely resisted the scheduled demolition. See John Sewell, *The Shape of the City: Toronto Struggles with Modern Planning* (Toronto, 1993).

47 See http://www.toronto.ca/culture/the_guild.htm [accessed 12 February 2012].

48 'The Guild Inn', *Carling Hospitality*, 3/7 (April–May 1949), pp. 4–6.

49 I borrowed the idea of the cemetery from a recent intellectual thriller by Spanish writer Carlos Ruiz Zafón, where he describes a 'cemetery of forgotten books'. Carlos Ruiz Zafón, *The Shadow of the Wind* (London, 2005).

50 Spencer Clark, 'The history of my life' unpublished biography. University of Waterloo Library, Breithaupt Hewetson Clark Collection, 1.2, box 6.

Index